WINNING THE GREEN NEW DEAL

WHY WE MUST, HOW WE CAN

EDITED BY **VARSHINI PRAKASH**
AND **GUIDO GIRGENTI** OF THE
SUNRISE MOVEMENT

SIMON & SCHUSTER PAPERBACK
NEW YORK LONDON TORONTO SYDNEY NEW DELHI

Simon & Schuster
1230 Avenue of the Americas
New York, NY 10020

First Simon & Schuster trade paperback edition August 2020

Interior design by Ruth Lee-Mui

Manufactured in the United States of America

1 3 5 7 9 10 8 6 4 2

Library of Congress Cataloging-in-Publication Data

Names: Prakash, Varshini, editor. | Girgenti, Guido, editor.
Title: Winning the green new deal : why we must, how we can / edited by Varshini Prakash & Guido
Girgenti of the Sunrise Movement.
Description: First Simon & Schuster trade paperback edition. | New York : Simon & Schuster, 2020. |
Includes bibliographical references and index.
Identifiers: LCCN 2020023589 (print) | LCCN 2020023590 (ebook) | ISBN 9781982142438 | ISBN
9781982142476 (paperback) | ISBN 9781982142483 (ebook)
Subjects: LCSH: Environmentalism—Political aspects—United States. | Green movement—United States. |
Environmental policy—United States. | United States—Environmental conditions.
Classification: LCC GE197 .W57 2020 (print) | LCC GE197 (ebook) | DDC 363.7/060973—dc23
LC record available at https://lccn.loc.gov/2020023589
LC ebook record available at https://lccn.loc.gov/2020023590

ISBN 978-1-9821-4243-8
ISBN 978-1-9821-4248-3 (ebook)

CONTENTS

INTRODUCTION

THE ADULTS IN THE ROOM

VARSHINI PRAKASH

I don't know how to say this except to just say it: *Young people have got to rise up.* That's it. That's the message. If you take one thing away from this book, let it be the courage to rise up alongside your friends and classmates and neighbors to demand a safe future for all of us, and the certainty that if we rise together, we will win.

I had that "rise up" feeling when I awoke early on September 20, 2019.

While I'd been asleep in the US, millions of young people around the world had been pouring out onto the streets of cities and towns small and large in unprecedented numbers. I scrolled through Twitter from my bed, awestruck at images of schoolchildren in Delhi chanting in unison, thousands in Japan, hundreds of thousands pouring out across Australia. Over 1.4 million people were already striking in Germany. In every country, thousands of youths gathered with homemade signs, images of the earth and words scrawled onto pieces of paper. My personal favorite: "If you won't be the adults, then we will."

I swung out of bed, ready to join tens of thousands more in New York City, where leaders like the indomitable sixteen-year-old Greta Thunberg would be rallying. I made my way to Foley Square, the sidewalks streaming with teenagers hours before the march was set to begin.

Groups of schoolchildren covered the green, some who couldn't have been older than five or six standing alongside teachers and parents holding banners. Thousands of teenagers were excused from class by New York City's Department of Education so they could strike. I caught up with activists like fourteen-year-old Alexandria Villaseñor and sixteen-year-old Jamie Margolin. The average age of the protesters was a full three decades lower than at any other mass demonstration I'd attended. The energy was palpable, buoyant, and fierce, heightened by the rising heat of the day.

I made my way to the stage, pushing through throngs of young people of all races, from every borough in the city, arriving backstage just as the first speakers bellowed into the microphone.

I looked out at the vast numbers reaching as far back as the eye could see. I'll tell you, there is nothing more beautiful than thousands of people standing as one, in active hope for a better future.

And yet, as I peered out, tears began flooding down my face, and I felt a tightness in my stomach. *These were children.* And they were compelled, instead of playing with their peers or studying or hanging out on a Friday afternoon and enjoying the last few moments of summer, to march, to fight, to demand, to plead for their futures. They knew irrevocable damage may be done long before they reach voting age or can run for office. They would be left in stewardship of a treacherous world, while older generations, dead and gone, would never face the consequences of their ruinous actions.

We are defined not by the conditions in which we find ourselves but in how we choose to respond. In this moment it was apparent that, against all odds, young people were choosing to fight back, young people were choosing hope, young people were choosing life. We are not the first generation to witness environmental disaster, but we are the last with the power to stop its cataclysmic force. We haven't caused this problem—nor do we know how our future will unfold. And yet we are choosing responsibility and action, choosing each other, over isolation, fear, and retreat. It's a scary thing, but it's all the more possible when we do it together.

All this rushed through my head when suddenly my name was called. I stepped up, the sun beating down, to a cheering crowd of 250,000 people. My strength came back, soaring into my lungs. and I jumped up onto the stage.

> When I first learned about the climate crisis as a kid, I would lie awake late into the night, heart pounding, imagining what it would mean for me and people who looked like me around the world. I couldn't get the images out of my head: of what people would do to each other when they were faced with no food or water, with cages and guns, when they sought sanctuary in other countries. I felt alone, small, powerless.
>
> Today, this generation is taking over. Our days of waiting for justice, our days of waiting for action, our days of waiting to be heard, are over—we'll put our feet to the streets until we get it done. Today kids don't have to feel small and alone and powerless. They can know they are part of the most important movement that's shaking our world from its roots.

That day, 7 million people went on strike.

This is the movement that is going to win a Green New Deal.

"WHAT IS YOUR PLAN?"

That "rise up" feeling comes in many different flavors. In September 2019, I was ready to rise from a place of joy and optimism. But less than a year prior, I was enraged.

On November 7, 2018, the morning after Election Day, I woke to the news that Nancy Pelosi, fresh off her victory in the midterms, would be using her new powers as Speaker of the House to create a committee to "educate the public about the impact of more frequent extreme weather events."

This was personal for me. Sunrise Movement, the youth movement of which I am a co-founder and the executive director, had been building toward this election since our foundation in mid-2017. We set out on a mission to make climate change matter in the midterms, as a key step toward our eventual goal of winning comprehensive policies to stop climate change while creating millions of good jobs.

Our election strategy focused on exposing both Democrats and Republicans for their financial ties to the fossil fuel executives profiting from climate change, while supporting those Democrats who would swear off fossil fuel money and back ambitious climate policy. Unlike some other climate groups, we wouldn't focus any energy on educating or persuading GOP candidates. The few climate-conscious GOP politicians had long since been expelled from office by the party's billionaire fossil fuel financiers, while the party as a whole displayed increasingly open contempt for people of color, democracy, and the bipartisanship that prior generations held dear. The GOP was simply too far gone.

That's why Sunrisers around the country had been busting our butts from June through November to register voters and promote candidates, all of them Democrats. Four million more young people turned out to vote than in the last midterm election, helping Democrats reclaim the House of Representatives. Our generation backed the Dems without hesitation—because of the obvious need to put a check on Donald Trump—but also without enthusiasm, because it was equally obvious that the Democrats still had no plan to actually solve the massive problems casting a shadow over our future. Sunrise's efforts had raised the profile of climate change in some races, but it was evident we still weren't where we needed to be.

And now Pelosi had confirmed that her climate strategy extended no further than "educate the public" through a series of stuffy congressional hearings that nobody would hear about. Personally, I thought the wildfires raging across California and blanketing the news were education enough. I was scared, angry, and ready to rise up.

Six days later, I walked with two hundred others into a Washington, DC, office building, to visit Pelosi in person. We flooded the halls with yellow and black signs, repeating a desperate message: "*We have 12 years. What Is Your Plan?*" We already knew the answer, of course—but I'll come back to that later.

I'd been to events like this before. I'd marched against the Keystone XL pipeline. I'd sat in the offices of my college administrators until they divested the school's endowments from fossil fuels. I'd been arrested working to shut down a polluting coal plant in southern Massachusetts.

I'd learned the hard way that victory usually doesn't come the same day that we sit in at a congressional office. Organizing to stop the climate crisis and secure jobs and justice for all Americans is daunting, in part because we are already starting so far behind from where we need to be, and we never win everything we need at once.

We watch our homes and our neighbors perish in the California wildfires; we get a call from our relatives across the world, telling us of the floods they hope to escape. We ask ourselves why these fires and floods keep getting worse, and our questions lead us to confront a fossil-fueled economy careening toward an unlivable future while politicians loudly deny the crisis and quietly ensure more coal and more oil is extracted and burned.

I've learned that it's not the job of an organizer to wallow in all the reasons we might fail. It is our job to remember that truth and justice demand we find a way to save as many lives and livelihoods as we can, to imagine what our country might look like if we took this emergency seriously, and then to build the power we need to make that a reality.

That's what I told myself when I woke up the morning of November 13. No matter what Rep. Nancy Pelosi does, our sit-in is another small tug at the arc of history. If our generation can step up and act like this crisis is an emergency, if we can force a conversation about the policies we need to avert catastrophe and protect our communities, then it will be worth it.

I stepped into the lobby of Nancy Pelosi's office and spoke.

> Our generation has witnessed the failure of political leadership as long as we've been alive on this planet. We are angry at the cowardice of our leaders at a time when courage is needed most.
>
> Wildfires are burning down city after city in California, and the latest UN report says we have twelve years to rapidly transform our economy and society to protect human civilization as we know it. We'll talk about it like that, 'cause that's what it is. Meanwhile, establishment Democrats take hundreds of thousands of dollars from fossil fuel executives and lobbyists and say they have no intention of putting forward a real climate plan before 2021. That's a death sentence for my generation.
>
> We've been told our whole lives to wait and see what the

> adults do. We're out here on day 1 of the new Congress because we don't have time to wait and "see" what our leaders will do.
>
> The last major *attempt* at climate policy was literally in 2009. That was when I was fourteen! Resting on our laurels won't bring back the thirty-one lives gone in the California fires. Resting on laurels won't save my family in India from drowning in floods. Resting on laurels sure as hell won't save the Democratic Party.
>
> Our generation just helped flip the House with a record turnout. We will no longer tolerate empty promises and words without action from the Democratic establishment. If Pelosi and the leadership don't step up, they need to step aside.

And then, famously, Representative-elect Alexandria Ocasio-Cortez entered the office and offered her support for our cause. Later that day, she released a proposal for something called the Green New Deal, a comprehensive program of reform to halt climate change while creating good jobs and reducing inequality.

In the days after that sit-in, it *did* feel like the arc of history had budged a little. AOC's Green New Deal proposal and our challenge to Pelosi sent a shock wave through the worlds of politics and climate advocacy. Some establishment politicians reacted with shock and anger that a newly elected congresswoman would challenge her own party so openly, while others scrambled to express support, lest they attract a sit-in in their own office.

Longtime climate advocates were amazed to see from us a holistic plan to provide a livable climate and good jobs to all Americans, black, white, and brown, all being discussed in the halls of Congress. This vision had been developed for decades by environmental justice and labor advocates who cautioned that to address climate change in isolation from racial or economic justice would be foolhardy. Now it was being discussed seriously not only in the halls of Congress but in the mainstream media. After decades of partial solutions and non-solutions, the political conversation was finally focused on change at the scale of the problem.

THE GREEN NEW DEAL

The Green New Deal is the reasonable, pragmatic response to a crisis that at this moment is almost unimaginably severe.

Critics have called it radical and far-fetched; such claims say more about the critics than about the proposals at hand. The Green New Deal only appears extreme when we, through ignorance or denial or forgetfulness, lose touch with the emergency we face. The house is on fire, but our would-be leaders are in the basement with headphones on. Will they hear the alarm before the ceiling falls in?

To stabilize our climate and preserve life on earth for future generations, we must rapidly halt carbon emissions. To do that, we must reform every aspect of our economy—the way we grow our food, get around our cities and towns, construct and heat and light our homes and offices.

The physical terrain of our lives will be rebuilt. Looking out my apartment window on Sumner Street, I imagine a Green New Deal for East Boston. Where now there is a maze of highway tunnels and traffic lights, we'll build an affordable bus system and light rail to take people where they need to go. Today the air is sour with the smell of exhaust and pollution; soon we will breathe easy. I can picture solar panels on the roofs across the street and wind turbines in the harbor. I love my corner bodega, but today I can't get more than cheap snacks there; soon it will carry fresh produce from a regional network of farms and gardens. We'll have a new seawall to help protect against rising seas, while some areas will have been restored to wetlands to provide a buffer against high tide. And so it will go, in its own way, in every single big city and small town in the country.

These changes are good and necessary, but they will also be disruptive to an economy long dependent on fossil fuels. There is risk of people being left behind in an economy shifting to meet the climate threat. People could get hurt, and the resulting political backlash could undermine the whole project. We saw this happen in France, where a fuel tax hike caused an economic crunch for the country's working class, and the ensuing popular backlash of the yellow vest movement not only sank the fuel tax but threatened the stability of the government itself.

That's why the Green New Deal recognizes the government's duty to guarantee fundamental human rights throughout our response to the climate crisis. It's the right thing to do, and it's the only way to get real climate policy implemented. The right to clean water, already violated in communities like Flint, will be made even more tenuous amid climate-induced drought, so water infrastructure must be at the top of our priority list. The right to quality health care through private employment could be imperiled as millions shift jobs due to climate policy; thus we must establish Medicare for All. The right to economic security can be met through a federal "job guarantee" policy offering dignified work to all who want it. And we must protect against institutional racism in all these programs, to guarantee that *all* people, black, brown, and white, benefit equally.

We'll ensure a fair and just transition for all workers in the fossil fuel sector and other energy-intensive industries. This transition will include guaranteed priority placement in new jobs for displaced workers with equivalent pay, early retirement options, and worker retraining programs. For the people who have risked their lives working in coal mines or on dangerous offshore drilling facilities to power our homes, it's the least our nation can do.

Rather than one priority among many, the Green New Deal is best thought of as a *governing agenda* that guides every aspect of public policymaking. Its many components will not all be accomplished through a single piece of legislation in Congress. Like the original New Deal, it will require dozens or hundreds of bills and executive actions, implemented over the course of a decade. The federal government must lead this project, in coordination with additional policies at the state level and in thousands of municipalities.

A governing agenda requires sustained *governing power.* For the Green New Deal to succeed, its backers cannot settle for one moment of victory before riding into the sunset. We must win and hold control of government for a decade or more. This kind of sustained, unapologetically progressive governance has only occurred twice before in this country's history—during the Reconstruction era following the Civil War, and during the New Deal era from 1933 through the end of the 1960s.

As someone growing up at the blurry edge of bipartisanship's death and the rise of political tribalism, all of this sounds unimaginable. I've watched politicians bicker and throw stones while temperatures, inequality, and white supremacy rise to the boiling point. One of the greatest political achievements of my lifetime was a modest improvement to our health-care system, and since then Congress has ground to a screeching halt. Forward-thinking governance is a daydream.

And yet it must be done. Doing it will require the power of organized people, organized institutions, and organized ideas. This book collects some of the key ideas about why we need the Green New Deal, what it is, and how we get there.

As we shape the Green New Deal together, we will experience the results in our everyday lives. I imagine dropping off my future children at public school, where the teachers who shape young minds will be compensated generously by a society that values the future. My children will learn the skills needed to thrive in the twenty-first century: in science, methods of ecological restoration; in the humanities, ways of articulating our fears and hopes for the future; in social studies, tools for conflict resolution and for understanding each other across our differences. On my way home, I see my neighbors putting up solar panels and a work crew building new affordable housing for displaced oceanfront residents. The neighborhood has several new murals, the product of a revived Works Progress Administration, the New Deal–era program that funded artists and cultural workers. Everywhere I see ordinary people doing honest work for the benefit of all.

I love thinking about what purpose all of us might discover in a Green New Deal America. There's a father in Brooklyn who was laid off from a design firm in the Great Recession. He keeps putting food on the table as a delivery truck driver; but he's got a green thumb and his favorite hobby is tending to the community garden down the block. There's a girl on the South Side of Chicago who suffers from asthma and wonders how she'll pay for college; she dreams of being a solar engineer so we don't have to burn coal any longer. There's a farmer in Iowa who is spraying pesticides on his corn and struggling to stay out of debt; he's interested in moving to more sustainable methods if he could only get a fair price for his crop. The Green New Deal is a chance for that father in Brooklyn, and that girl in Chicago, and that farmer in Iowa, to do the work they love and the work that helps people, and to be rewarded for it.

By providing good work with family-sustaining wages, the Green New Deal offers freedom from the drudgery and stress of having to work multiple mind-numbing jobs just to make ends meet. And in focusing our collective attention on the common good rather than selfish individual advancement, we may rediscover the secrets of what makes a good life. Here in the apartment on Sumner Street, family and friends and colleagues will break bread together around a long table, drink wine, and enjoy each other's company, not on a special occasion but as a regular practice. The Green New Deal is good living.

When disaster strikes, as we know it will, we'll be prepared. We'll respond in mourning, but with the consolation of knowing that we are all in it together. Access to relief services will be determined not by individual power and privilege but by collective need.

I hear clamorous voices proclaiming that none of this can be done. Across the corporate-owned airwaves and in stiff DC press conferences, they insist that survival is nothing more than a "green dream." In response to all, we humbly, quietly insist again: "But it must be done."

If you are looking for reasons to doubt, you will find no shortage of them. The cynic will always find plenty of evidence. I am not here to persuade you that things are not bad, or that the needed changes are likely. I am simply asking you to embrace the work of making the impossible possible. Greta Thunberg has said: "When we start to act, hope is everywhere. So instead of looking for hope—look for action. Then the hope will come." I have dedicated my life to action, and I find myself paradoxically full of hope and brimming with sorrow all at once. I'll take it over the alternative of dull numbness or fatalism. It feels like being alive.

There are much, much, much worse alternatives to a Green New Deal. I don't just mean climate catastrophe. I mean the way we respond to it. There are ways that our society may adapt to cities going underwater, to landscapes being permanently erased due to drought and wildfire, to an influx of climate refugees, that are haphazard, shaped by the evils of racism and wealth inequality, and that leave people to make it on their own. This is the path we take if we do nothing.

The Green New Deal is a different choice: to survive together rather than perish separately. It is a rejection of the bunker mentality—the Wall mentality—that fearful and unimaginative men seek to impose upon the populace. It is a commitment to fight for somebody we don't know. Yes, this requires a leap of faith. It requires us to believe in each other.

When I am with other young people, I believe. I believe that we will not be bowed by life's defeats. I believe that we can learn to heal together. I believe that we can discover the good life together. I believe that we can remake America, that we can be the America that never was. This is the project of our lives. The time to begin is right now.

LET'S RISE UP TOGETHER

The year and a half since that day in Pelosi's office has been the wildest time of my short life. As a spokesperson for and representative of our growing movement, I've been invited to meet with congressional leaders, media figures, and corporate and nonprofit executives to share our message. Sitting at those mahogany conference tables and five-course dinners, surrounded by some of the most powerful people in the country, I've learned without a doubt the answer to our original question: *"What Is Your Plan?"*

The answer? They have no plan. They don't even have a plan to make a plan.

Let's be more precise: they have no idea what the fuck is happening or what to do about it. The power elite has fallen asleep at the wheel.

This is why young people have got to rise up.

After every discouraging meeting with a suit, I am blessed to return to a movement full of young people who are looking to the future with clear eyes and open hearts. When I am with Sunrisers, I believe that a better world is possible. In the face of a million people insisting that it can't be done, in the face of our own anxieties that we aren't good enough or smart enough, that we are destined to witness the unraveling of everything we know and love, the young people of this movement wake daily and set their spirits to the project of collective liberation.

Sunrise is deeply indebted to a much larger movement of which we are a small part. We haven't spent centuries preserving ancestral wisdom of how to live in the right relationship with the land—Indigenous peoples did that. We didn't invent the idea of a just transition that takes care of energy industry workers amid the transition to a new economy—workers who organized through the union movement did that. We didn't spend decades sharpening and sharing a true analysis of the deep and inextricable connections between racism, our economic system, and climate change—the environmental justice movement did that. We haven't inspired the largest numbers of people to take to the streets in recent years—Greta Thunberg and the Friday school strikers did that.

Sunrise's unique contribution, where we have excelled, has been in taking the fight to the political arena, bringing some of these ideas to greater prominence in the mainstream. We've done this by combining a political savvy with a disciplined and fast-growing movement set on a clear goal. This has been our unique contribution so far. We're honored to have many of the thinkers who have shaped our approach to politics and movement-building represented in this book.

Whether you're part of Sunrise, or one of our allies, or not yet engaged, we hope this book can be a tool to help us all rise up together.

EDITORS' NOTE

The seeds of this book were planted over the past decade, as student activists—the core of what would become Sunrise—faced the fierce urgency of the climate crisis and wrestled with questions of organizing and movement-building. *What solutions are big enough, transformative enough, to meet the crisis facing humanity at this moment in our history? How do we build a movement big enough and strong enough to win?* These are questions of vision and strategy, theory and practice.

With this book, we hope to provide the reader with some of the answers we've found on our journey, a vision for human prosperity, and a strategy for how we might realize this vision in our lifetimes. This is the book we wish we could have read five years ago. It is not a comprehensive or definitive statement on the movement for a Green New Deal. It is a compass for people navigating the perils and possibilities of this moment. We hope that the victories we win in the next decade make the ideas discussed in the following pages appear, in retrospect, to lack ambition.

The book's three sections answer three questions about the Green New Deal (GND): why the crises we face demand a solution with the GND's scale, speed, and commitment to economic and racial justice; what the GND's guiding principles and policies are; and how we can organize to win a GND in this coming decade. The sections are not chronological. Many of our movement's strategies were developed long before senators and representatives were proposing Green New Deal bills; many of the principles that inform the Green New Deal's policies were developed in the struggles for environmental, racial, and economic justice over the past half century, long before the United Nations Intergovernmental Panel on Climate Change (IPCC) called for unprecedented social and economic transformations to avert disastrous warming.

This book is tied to a specific moment—a moment when winning a Green New Deal seems not only necessary but possible. It also humbly follows centuries of struggle for a multiracial democracy that guarantees freedom and justice for all who call themselves American, guided by the thinkers and organizers—some of whom authored chapters in this book—who've made the Green New Deal possible by carrying the torch for many decades before us.

We hope that this book provides some light to those who will carry the torch alongside us and beyond us.

VARSHINI PRAKASH and GUIDO GIRGENTI

PART I

THE CRISIS THEY WON'T LET US SOLVE

ONE

THE CRISIS HERE AND NOW

DAVID WALLACE-WELLS

How bad is it? Almost certainly worse than you think. The climate crisis is no longer a story to be foretold in the future tense, but one unfolding, catastrophically, in the present. It can no longer be defeated, only restrained. And human suffering on a scale that once seemed unthinkable, even to alarmists, has arrived unmistakably on the horizon, though we often fail to see it clearly.

We are already living in unprecedented times. The planet has warmed about 1.1 degrees Celsius since the beginning of the Industrial Revolution. While that may not sound like much, it already puts us entirely outside the window of temperatures that enclose all of recorded human history. Which means that everything we have ever known as a species is the result of climate conditions we have already left behind. It is as though we have landed on a different planet, and now have to determine what parts of the civilization we've smuggled with us can survive, what needs to be reformed, and what must be discarded.

How different is our new planet? Since the 1970s, the area of land burned by wildfires in the American West has doubled, and the number of large fires has quintupled; as much as 60 percent of animal populations on earth have died since then, the result of ecosystem loss and pollution and warming, as have perhaps 70 percent of insects; nations in the global South have lost as much as a quarter of their potential GDP, and more than 9 million people are dying each year, already, from the air pollution produced from the burning of fossil fuels—an annual death toll the equivalent of the Holocaust.

Things will get worse from here—probably a lot worse. But a terrifying future shouldn't distract us from a horrific present: the Greenland ice sheet melting seven times faster than just a few decades ago, European heat waves testing temperature records three times in a single summer, and Houston hit by five "500-year storms" in the last five years. Five centuries ago, Hernán Cortés had just landed in Mexico and there were no European settlements yet in North America, which means this is a storm that should be expected only once during that entire history—the arrival of European colonialists, the waging of a genocide against the continent's native peoples, the fighting of a revolution and the establishment of an empire of slavery, the fighting of a civil war, industrialization, the waging of two world wars, the era of American empire and the Cold War, the end of the Cold War and the "end of history," September 11 and the Great Recession.

One such storm was expected in all that time. Now five hit within a half decade. This is the world we live in today, already lethal and brutal and yet a better-than-best-case scenario for climate change. Warming will not stop tomorrow, and the emissions of yesterday and today and the next decade will, if they continue at anything like the current pace, make things much worse.

This is why, the science says, the choices made in the next decade will define the shape not just of the near future but perhaps of many centuries to come—in this way, climate change is not just incredibly rapid but unfathomably long, the effects we produce today defining life on earth for perhaps as long as humans are around to witness and record it.

In 2018, the UN IPCC released an eye-opening report declaring that "rapid and far-reaching transitions . . . unprecedented in terms of scale" would be necessary to keep warming below 1.5 degrees, the threshold nations of the world agreed to target in the Paris accord. To meet that target, the IPCC announced, would probably mean roughly halving global emissions—which are still rising—by 2030. That would require, the report of the world's scientists suggested, a global mobilization against climate change like that which the United States undertook in World War II—when nearly every man of fighting age was drafted into the military and nearly every woman of working age into the workforce, when factories were repurposed and entire industrial sectors nationalized in the space of months.

Past 2030, things wouldn't get easier: the planet would need to entirely zero out on carbon emissions by roughly 2050, the IPCC said. Technically, that would be necessary to stabilize the planet's temperature at any level of warming, even a hellish one, since just a sliver of the carbon emissions being produced today will always mean some additional heating of the climate. At today's rates, we would entirely exhaust the carbon budget that would allow us to limit warming to 1.5 degrees in less than a decade.

Erik Solheim, then the head of the United Nations Environment Programme, described the message of the report as "a deafening, piercing smoke alarm going off in the kitchen." And indeed, the public was alarmed. The report was followed by a surge of political mobilization around climate change, with activists and disengaged citizens alike terrified by how short the runway seemed: twelve years to halve emissions. That does not mean 2030 is an expiration date, a suddenly arriving apocalypse beyond which decarbonization is pointless. But that fact should be no comfort. The more warming we bake in today, the harder anything we might want to do down the road will be. And there is absolutely no path to staying below 1.5 or 2 degrees of warming that does not involve a society-wide plan for decarbonization in the next decade.

How the world could get to the goal is very much an open question, but if we hope to pursue it in any way deserving of the word "just," the wealthy nations of the world, who have benefited most from warming, must lead. To get to net-zero emissions by 2050, while allowing developing nations to decarbonize, the United States—the world's most powerful economy, with the largest historical share of carbon emissions—will need to cut emissions much faster and much sooner. Other nations—particularly rapidly developing nations like China, India, Brazil, and others—have little incentive to act rapidly and decisively if we are not doing the same. Many analyses suggest that a "fair share" of US action would have us decarbonizing over the next decade while also supporting the decarbonization of other countries.

If we fail to halt and ramp down emissions this decade, we will reach 2 degrees of warming between 2040 and 2050. The difference between 1.5 degrees and 2 degrees may not seem like much, but with climate change every tiny differential comes with an enormous, if not unprecedented, amount of human suffering. At 2 degrees, 153 million more people would die of air pollution than in a world warmed by 1.5 degrees. On that hotter planet, flooding that today hit once a century would arrive every year, the UN expects. Many of the biggest cities in South Asia and the Middle East will be unlivable in summer. These are cities that today are home to 10, 12, 15 million people, and, as soon as 2040 or 2050, you won't be able to walk around outside in them, and certainly wouldn't be able to work outside in them, without risking heatstroke and possibly death on certain summer days.

This is one reason the UN expects—just by midcentury, at about 2 degrees of warming—that the world could see 200 million or more climate refugees. By that time, changes to the oceans alone could displace as many as 280 million people. Wildfires in the American West could be four times worse than they are today—four times worse than fires that regularly burn millions of acres every year; beyond 2050, scientists are reluctant to even make predictions. So totally burned will the region be, they say, that we don't know what new plant life will grow up again amid the ashes, and so we can't know at what rates or under what conditions that new life will burn. And at just north of 2 degrees, we will probably lock in the permanent and irreversible loss of all the planet's ice sheets—enough to raise the global sea level, over centuries, more than 200 feet. That would flood two-thirds of the world's major cities.

The scientists of the world call 2 degrees of warming "catastrophic." The island nations of the world call it "genocide" and chant "1.5 to stay alive" at international climate negotiations. It is the future that Green New Dealers strive to avoid, by demanding that the US and other industrialized nations muster and mobilize all of our resources to do everything in our power to get emissions as close to net-zero as possible in the next decade. And yet, the focus on these thresholds—1.5, 2 degrees—has perhaps implied that this is the full range of warming we are at risk of. That 2 degrees is the worst we might experience. Given the obstacles facing climate action, and how rapidly decarbonization would be needed to avoid it, 2 degrees is, practically speaking, more of a best-case scenario.

We are not on the path to 2 degrees; we are nowhere near that path. If we halted global emissions tomorrow, we would probably be due for about a half-degree more of warming, just from the carbon in the atmosphere already—bringing us to about 1.5 degrees. Different models project different temperature increases for the end of the century, mostly because the biggest variable is human action: how we respond, and how quickly. But without very aggressive decarbonization, the planet is nearly certain to warm by at least 3 degrees Celsius by the end of the century.

In its last major report, the United Nations (UN) projected that the path we're on today could take us north of 4 degrees, and a new generation of more sophisticated models suggests the warming could be higher still. And at 4 degrees, it's been estimated that global GDP could be as much as 30 percent smaller than it would be without climate change—an impact twice as deep as the Great Depression, and permanent. Climate damages could reach $600 trillion—hundreds of trillions more than all the wealth that exists in the world today. Megadroughts could produce regular "multiple breadbasket failures," producing widespread famine. According to some estimates, whole regions of Africa and Australia and the United States, parts of South America north of Patagonia, and Asia south of Siberia could be rendered uninhabitable by direct heat, desertification, and flooding. Certainly it would make them inhospitable.

We could have more than twice as much war, because of the relationship between temperature and conflict, and half as much food, because agricultural yields decline as temperatures rise. Parts of the planet could be hit by six climate-driven natural disasters at once, which, among other things, calls into question the ability of any government, no matter how wealthy and well run, to respond. Climate scientist Kevin Anderson has said that 4 degrees of warming is a world "incompatible with an organized global community" and "is likely to be beyond 'adaptation.'" Projections like that lie beyond the realm of science, strictly speaking. But so does our future: we simply cannot know in detail what things will look like at these temperature levels, they are so far from all of human experience.

And yet our lives will all be defined by those changes, no matter who we are or where we live. Nothing, and no one, lives outside of the crisis, which is all-encompassing, all-touching, all-transforming already. This is the perverse promise of warming: nearly everything that is broken or brutal about the world today will be made worse by climate change, should we not take action to stop it. And nearly everything that is alluring and vibrant and vital can be made better and brighter, too, through decarbonization, should we pursue it.

We might assume that a crisis of this scope took centuries of slow, steady boiling to reach, beginning when the first carbon was burned in eighteenth-century England. But in fact, more than half of all the carbon exhaled into the atmosphere by the burning of fossil fuels has been emitted in just the past three decades. Which means we have done as much damage to the fate of the planet and its ability to sustain human life and civilization since Al Gore published his first book on climate than in all the centuries—all the millennia—that came before. Eighty-five percent of all fossil burning has happened since the end of World War II, meaning the story of the industrialized world's kamikaze mission is the story of a single lifetime.

The UN established its climate change framework in 1992, advertising scientific consensus unmistakably to the world; this means we have now engineered as much ruin knowingly as we ever managed in ignorance. And we are doing it very rapidly, producing each year more carbon emissions than we did, collectively, the previous year—not just by not moving quickly enough to decarbonize, in other words, but moving decisively in the wrong direction, even today, accelerating into an irreversible cataclysm every year. Today, we are adding carbon to the atmosphere ten times faster than the rate at which greenhouse gases filled the atmosphere during the worst mass extinction in the earth's history, when perhaps as much as 97 percent of all life on earth died.

This is bad news—but it contains, at least, a silver lining of political opportunity. Contemplating the worst-case scenarios—twice as much war, half as much food, a permanent economic depression, a crisis "beyond adaptation"—can feel paralyzing, because the scale of those impacts is so large. But they are ultimately a reflection of just how much power we collectively retain over the climate. If we bring about those catastrophic scenarios, it will be because of actions humans will take from this point forward. Which means we can take a different kind of action as well, and produce just as dramatic an effect in the opposite direction: toward a just and prosperous and fulfilling green world.

That may seem naive. And the obstacles are, indeed, enormous. But it is also a simple fact that the main driver of warming is human action—how much carbon our economies put into the atmosphere. At some point, should we not change course, we will eventually trigger feedback loops by which the climate system warms itself, perhaps quite rapidly and dramatically. But for now—and for the pivotal decade we are entering—human hands are on those levers, and we are deciding, as a society, what kind of future we want for ourselves, our children, and our grandchildren.

All of which raises the question: Who is "we"? Punishments from warming are already unequal, and will grow more so—within communities, within nations, and globally. Those with power to make meaningful change are often those today most protected from warming; in many cases, they are those who stand to benefit, sometimes handsomely, from inaction. But those who stand to suffer from that inaction, and those disasters, number many times more. All told, the cost of changing course is now understood to be so much smaller than the cost of standing still that only a stranglehold on our political economy could explain the indifference of public policy toward the existential challenge we face, collectively, in the form of extreme weather and natural disasters and generalized climate suffering.

That is not to say making a different choice will be easy. We are now burning 60 percent more coal than we were just in the year 2000. And energy is, actually, just one part of it. There is also the need to get to zero emissions from all other sources—deforestation, agriculture, infrastructure, livestock, landfills. And the need to protect all human systems from the coming onslaught of natural disasters and extreme weather. And the need to erect a system of global cooperation, to coordinate such a project.

The speed and scope of this crisis make it hard to comprehend all at once; on the one hand, we've brought our world to the brink of catastrophe in the span of a lifetime, and we are determining now, in this next decade, whether we keep our world within the bounds of comfortable habitability. At the same time, we've already transported ourselves to a planet our civilization was not built for, and we're already locked into decades of disruption and deterioration that will leave no part of our world unchanged. What that means is that we have not at all arrived at a new equilibrium. It is more like we've taken one step out on the plank off a pirate ship.

Perhaps because of the exhausting false debate about whether climate change is "real," too many of us have developed a misleading impression that its effects are binary. But global warming is not "yes" or "no," nor is it "today's weather forever" or "doomsday tomorrow." It is a cascade of cruelties, ruptures, and unravelings that gets worse as long as we continue to produce greenhouse gas. And so the experience of life in a climate transformed by human activity is not just a matter of stepping from one stable ecosystem into another, somewhat worse one. The effects will grow and build as the planet continues to warm. If we are already surrounded by what can seem at times like an unimaginable world, then we must be spurred to faster action to avoid even more unimaginable ones arriving, and soon. The future is very literally up to us.

“Paradise—may it be all its name implies”

Mikala Butson

In Paradise, California, there used to be a sign that read “Paradise—may it be all its name implies.” I grew up surrounded by trailer parks, decrepit mobile homes, and unkempt yards. While it was not always, in fact, a place that was all its name implied, Paradise was still my town. Now, people know it as the home of one of the deadliest and most destructive wildfires in history.

On November 8, 2018, my mother sent me photos of the fire. She had been working her early-morning job when she spotted a strange-looking sunrise peeking out over the mountains in the direction of Paradise. My brother was at home asleep. She raced back toward the fire to get him. As she drove, flames traversed the mountainside and black smoke engulfed the trees.

I remember my mom calling me and asking me if I wanted her to grab anything from my room.

I thought about the home I had known for as long as I could remember. I thought of the tree in the front yard where our first dog was buried. I remembered getting ready for prom and dancing around my room after ballet classes. I reminisced about playing make-believe games with my brother. Those memories could not be taken along in a car.

I told her I did not need anything—all I wanted was for my family to be okay.

Then, silence. We lost contact.

I frantically searched every Facebook group I knew of looking for answers. I saw videos of people driving through flames and burned bodies trapped in cars. I read stories about people stalled in parking lots and sheltering in pharmacies. I didn't know where my family was. All I knew was that they were driving through Paradise where the fire was growing by a football field every second.

I called nonstop with no response.

Several hours later, my brother called. He and my mom had escaped. They said they would have gotten caught in the fire if it wasn't for a man frantically waving and screaming at them to turn around and drive the other way. They told me about the elderly people they saw walking with their suitcases because they didn't have cars of their own.

Our home sustained extensive smoke damage. The town I knew for nearly my entire life now consists of rubble, contaminated drinking water, and dangerous debris. We fought for months with the insurance company, eventually selling our home as is and leaving nearly everything behind because we couldn't afford a moving truck.

I wanted to do something but I felt there was nothing left to fight for. Then, I found Sunrise.

I was invited to tell Paradise's story and since then I haven't stopped fighting. Every day, I think about those burned places—not just because they were places I enjoyed but because they were owned by people. People whose whereabouts remain a mystery to me. I think about how my story can, and will, become the story of many others—of towns that will burn or drown or otherwise collapse under the impact of climate change. For them, I will fight until every town is guaranteed a just and sustainable future.

TWO

WE DIDN'T START THE FIRE

KATE ARONOFF

It's easy to feel hopeless about the climate crisis hurtling toward us, threatening 500-year floods, thawed-out ancient plagues, and other disasters seemingly plucked from the Old Testament and plopped into our ten-day weather forecasts. The full scope of this destruction is almost impossible to comprehend. To cope and to understand, we break it down into parts we can wrap our heads around, to some unseasonably warm day in mid-October or whatever storm crashes into our city. We also filter it through whatever ideas—about politics or economics—happen to be lying around.

To many of us living in the global North, at least, those ideas don't offer much help in understanding the problem—let alone what to do about it. For nearly half a century now, an extensive network of right-wing think tanks, media outlets, and politicians have attempted to infuse every public policy debate under the sun with a warped moral philosophy: if you're in trouble, it must be your own fault. And you alone can fix it. "Too many people," former British prime minister Margaret Thatcher said, "have been given to understand that when they have a problem it is government's job to cope with it . . . They are casting their problems on society. And, you know, there is no such thing as society. There are individual men and women and there are families. And no governments can do anything except through people, and people must look to themselves first."

Her counterpart across the Atlantic, Ronald Reagan, echoed that sentiment, warning of what he called the nine most terrifying words in the English language: "I'm from the government and I'm here to help." Nearly half a century on, these mantras still shape the way we look at the climate crisis. Faced with the greatest existential threat the world has ever faced, they suggest that there is no such thing as an organized, society-wide response, and no meaningful role for the government to play. There are only individuals, each of us is equally to blame, and each of us must do our small part in isolation, becoming more conscious consumers and sacrificing for the greater good.

After years of denial and delay by the most powerful people on earth, then, Thatcher's words still hold sway. Earth Day listicles tell us how to lower our carbon footprints by having fewer children and eating less meat as oil executives board private planes to go stake out their newfound reserves. It's not only that this is a problem of our own making, either. It's one we're allegedly hardwired not to fix, and the only choice now is to embrace the inevitable apocalypse we led ourselves into.

Consider the folk wisdom of novelist Jonathan Franzen. "Call me a pessimist or call me a humanist, but I don't see human nature fundamentally changing anytime soon. I can run ten thousand scenarios through my model, and in not one of them do I see the two-degree target being met," he mourned recently, referring to the temperature threshold inscribed in the Paris Agreement. Detailing an earlier round of climate talks, writer Nathaniel Rich offered a similarly dreary take on human nature for the *New York Times Magazine*. "We have trained ourselves, whether culturally or evolutionarily, to obsess over the present," he wrote, adding that we "worry about the medium term and cast the long term out of our minds, as we might spit out a poison."

But who, in this story, is we? Laying the blame on all of humanity, and on our damned, irredeemable nature, doesn't make much sense when it comes to the climate crisis. There are staggering inequalities between wealthy nations like the United States—the world's largest historical emitter of fossil fuels—and the poorer countries already being worst hit. Within countries there are vast differences too, lest Charles Koch's carbon footprint stand in for that of all Americans. The crisis will, of course, affect all of us at some point, but there is no universal *we* fueling rising temperatures or blocking attempts to cap them.

Simply put, we don't all have our hands equally on the levers that fuel climate chaos. A tiny set of very powerful corporations and oligarchs spent prodigious amounts of money to keep burning and extracting toward destruction. For years they lied to us and told us that there's no crisis. After losing that fight, they spent almost as much money convincing us to blame ourselves for it, conveniently building on Reagan's doctrine of individual responsibility. It was none other than BP—the world's sixth-largest polluter—that popularized the concept of a personal carbon footprint.

All that cloying rhetoric about individual responsibility has just been the public face of a push to peel back regulations and protect corporate profits from such grave threats as taxes and democracy, taking resources from the many and redistributing them among the few. To draw attention away from the incredible solidarity found at the top of the income and wealth bracket, it pits working people in competition with one another instead of against their bosses. And it's worked wonders. The concentration of wealth at the top in the United States has returned to levels not seen since the 1920s, and the richest 400 families here—the top 0.01 percent—now pay a lower tax rate than the bottom 50 percent of households. These asymmetries track closely to greenhouse gas emissions: the wealthiest 10 percent of the world's population produce about half of its greenhouse gas emissions, while only about 10 percent of emissions are produced by the poorest 50 percent, concentrated—like climate impacts, cruelly—in the global South.

But to really get a sense for how unequal carbon footprints are, we have to look at corporations. Since the dawn of the industrial age, just ninety corporations—most of them fossil fuel producers—have produced two-thirds of planet-heating emissions. Since 1965, just twenty shareholder and state-owned fossil fuel companies have been responsible for 35 percent of the world's energy-related carbon dioxide and methane emissions. Accounting for Rex Tillerson's $145 million in ExxonMobil shares circa 2015—and the emissions that company creates—researcher Dario Kenner found that the CEO was responsible for over 52,000 metric tons of carbon dioxide that year: well over 3,200 times that of the average American, 6,400 times the average resident of China, and 38,400 times the average Indian.

One might argue—as executives like Tillerson do—that these corporations are simply producing their supply to meet our demand. That argument, though, ignores how central they've been to keeping that demand high and growing it, despite every shred of evidence suggesting that their business model—to dig up and burn as many fossil fuels as possible—is driving the world off a climate cliff. For decades, ExxonMobil and Shell misled the public about the fact that temperatures were rising and that their products had plenty to do with it, publicly refuting conclusions reached by their own scientists internally. In the US, ExxonMobil funded climate denier groups like the Heartland Institute to blanket state legislatures with white papers of junk science and gin up a debate in the mainstream media about climate science that never would have existed otherwise.

They've also helped transform the Republican Party into the political arm of the fossil fuel industry and bought off plenty of Democratic politicians as well. From 1989 until 2002, ExxonMobil, Shell, and more than forty other corporations and trade associations for polluting industries used the innocuously named Global Climate Coalition to undermine the Intergovernmental Panel on Climate Change and lobby against US involvement in the Kyoto Protocol, a binding UN climate pact reached almost twenty years before the comparatively watered-down Paris Agreement, before many Sunrise organizers and youth climate strikers were born. The same fossil fuel companies continue spending billions of dollars lobbying policymakers against commonsense regulations and to use federal law to open up new sources of and markets for their products that are cooking us, raking in an estimated $5.1 trillion in direct and indirect subsidies worldwide.

After decades of disinformation, it's time to place the blame where it belongs and design policy accordingly. We didn't start the fire, and the fire's burning because a clique of incredibly wealthy corporations does everything in their power to keep digging up coal, oil, and gas. The question for us now is not whether we can turn off the lights earlier or change "human nature." The question is whether we—our societies, our governments, and our economies—can execute a massive transition off fossil fuels and build a greener, fairer world.

If we are serious about that transformation—about survival—then what will it actually take?

The IPCC's 2018 report found that capping warming at 1.5 degrees Celsius "would require rapid and far-reaching transitions in energy, land, urban and infrastructure (including transport and buildings), and industrial systems," changes that "are unprecedented in terms of scale . . . and imply deep emissions reductions in all sectors." In 2019, the UN Environment Programme called for global emissions reductions of 7.6 percent *every year* between 2020 and 2030 to meet the same goal. The greatest recorded annual drop in emissions to date—of 5 percent—was triggered by the collapse of the Soviet Union, when one of the world's largest and most polluting economies ground to a virtual halt. Thousands of miles' worth of farmland were abandoned and in short order began soaking up colossal amounts of carbon dioxide.

Without the wide-scale deployment of technologies to capture and store carbon dioxide from fossil fuel plants and the atmosphere—which may or may not even be feasible—global coal, oil, and natural gas usage will, by 2050, have to decline by 97, 87, and 74 percent, respectively, per the IPCC, to cap warming at 1.5 degrees Celsius. That decline ought to happen fastest in places like the US, whose centuries of fossil-fueled economic growth enable bringing massive amounts of renewable energy online more quickly than in poorer countries. The stubborn, well-documented link between rising economic growth and rising greenhouse gas emissions also means that emissions are likely to rise outside the global North as those economies grow and people consume more. To allow space for that to happen—that is, to avoid imposing some draconian limit on new consumption in the poorest parts of the world—emissions in wealthier countries should reach net-zero by around 2030, as they provide vitally needed financial and technical support to make low-carbon development possible.

In other words, to avoid runaway catastrophe, we must overhaul our energy, food, and transit systems, on an unprecedented scale and timeline, to achieve emissions reductions never before seen in the history of humanity. There is no guarantee we will succeed. If we do, it'll be thanks to a collective effort several orders of magnitude greater than anything humanity has accomplished before. The hard truth is that no one individual can wake up and decide to build a mass transit system, electrify the economy, or phase out fossil fuel extraction. So long as we continue to live in an energy and economic system structured root to branch around fossil fuels, there is no accumulation of voluntary lifestyle choices that can accomplish these lofty goals. If we pull off this transformation, it'll be because big, activist governments set their sights squarely on constructing a carbon-free economy as fast as possible. The Green New Deal is from the government, and it's here to help.

If we look at the needed transitions, sector by sector, we get a sense of their scale. For starters, we need to triple the amount of economic activity that takes place on the electricity grid, which currently does not provide the energy to heat and cool our homes and start our cars. That means putting millions of people to work building thousands of miles' worth of transmission lines and overhauling the power sector as we know it, quite literally rewiring the country so that utilities can accept electricity from homes and communities instead of just distributing it.

Our suburban-centric urban sprawl will have to be filled in with lush networks of public transit and dense, affordable housing as combustion engines become a relic of the past. Much more of our food will need to be grown close to the places we eat it, and millions of people will need to be relocated as low-lying cities are battered and swallowed by the stronger storms and sea level rise already heading our way. We'll need to figure out how to take an awful lot of carbon dioxide out of the air, too, with the surest strategy being to plant billions or even trillions of trees. Sectors that are more difficult to decarbonize, like aviation and construction, are still in dire need of trillions of dollars in research and development (R&D) funding. This is all just the tip of the iceberg.

On the energy front, avoiding outright global catastrophe requires both that we build out massive amounts of clean power infrastructure and keep fossil fuels in the ground by challenging the business model of the coal, oil, and gas industry head-on. As it tackles energy consumption, a Green New Deal also needs to transform consumption in ways that curb energy demand, making low-carbon lives available to all. Those massive investments in public transit and energy-efficient affordable housing would go a long way toward decreasing the carbon footprint of our cities and towns and making them altogether more livable. A four-day workweek would leave more time for low-carbon dinners with friends and trips to the beach. National fuel-efficiency and renewable standards, like those in place at the state level in California, could require utilities and car companies to ramp up clean power usage or face harsh penalties. And a Cash for Clunkers program could mean that those who do still have to drive wouldn't be spewing greenhouse gases into the air while doing so.

All this could be a boon to clean energy companies and several other industries that will be essential to the transition. However, much of the work required for such a massive and rapid transition simply isn't profitable enough—in that way, it's like teaching and firefighting. It will be a stubbornly tough sell to risk-averse private investors, pointing again to the need for a strong public sector acting in the interest of the common good.

Simply adding more renewables to the grid won't be enough to drive down fossil fuel use to the extent needed, either. We've got to cut off fossil fuel pollution at its source, by taming the corporations hell-bent on extracting coal, oil, and so on. Even as solar and wind power have gotten cheaper and more widespread in the last several years, the share of energy they generate nationwide has remained largely flat. The rate of fossil fuel extraction has continued to increase and shows few signs of slowing down—particularly as the United States becomes a net exporter. In other words, the good isn't out-competing the bad. Placing real constraints on coal, oil, and gas companies is the only way to ensure we can reach our climate goals.

One place to start is ending the roughly $20 billion in subsidies the government hands over to the fossil fuel industry each year at the state and federal levels, including long-standing tax giveaways. (Permanent tax breaks to the fossil fuel sector are seven times greater than those awarded to renewables.) In banning new offshore drilling exploration, the United States would follow the example already set by New Zealand. And a Green New Deal could move to end the following practices: leasing federal land and water to drillers; granting permits for the kind of new infrastructure that allows companies to export oil and gas abroad; and reinstating the crude oil export ban, lifted in 2015. Handouts to the fossil fuel industry are so extensive that the Stockholm Environment Institute has estimated that as much as half of new oil and gas development would be unprofitable without them. Removing the perverse incentives that keep them afloat should be a no-brainer.

But more dramatic action may be needed; time and again the industry has shown that it will fight even light regulations with well-funded disinformation and armies of lobbyists. Exxon, Chevron, and their ilk will not politely relinquish hundreds of billions of dollars in fossil fuel reserves. In its quest to dismantle the public programs of the New Deal, the Right—in lockstep with corporate America—has yelled "Stalinism!" at anyone who suggests the government wield its power to steer the economy in the public interest, all the while lobbying the government to give corporations lavish tax breaks and legalize the conversion of public lands and public programs into extractive moneymaking schemes. Let's not forget that Wall Street has been this century's greatest beneficiary of government largesse.

To hasten the transition to net-zero emissions, researchers at the Next System Project have suggested that the US government could bring US-based fossil fuel industries under public ownership by buying up 51 percent of their shares—investors could sell these voluntarily, or else be made to compulsorily—then moving swiftly to curtail production, kneecapping these companies' capacity to pollute our political system. Of course, cutting off new extraction overnight would leave workers and communities that depend on fossil fuel employment in a lurch, and cause arguably catastrophic ripple effects throughout the US economy if alternatives were not at the ready. Whatever ownership structure is in place, living up to the values embodied in a Green New Deal will require a managed decline of the fossil fuel industry, in which the first priority is ensuring a dignified quality of life for extractive sector employees and the communities they live in—not golden parachutes for C-suite executives.

As we debate the policies of the coming transition, we can expect to hear fossil fuel executives supporting a modest carbon tax as an alternative to Green New Deal–style policymaking. They know all too well that subtle price signals cannot wholly decarbonize the economy on the timeline needed. The math for them is straightforward: a modest tax on carbon will finally kill coal while driving companies toward more oil and gas production that the tax is too low to deter. Combined with a few token investments in low-carbon fuels, it'll also pacify critics and ward off more stringent regulations.

More likely is that a meaningfully strong carbon tax won't pass at all, and business as usual can keep humming along. While one industry-backed plan would start at $40 per ton, the IPCC report referenced above says that a *global* carbon tax ranging from $135 to $5,500 per ton would be needed to avert a catastrophic rise in temperatures. For reference, Sweden has the highest carbon tax on earth at $123 per ton, and existing prices on carbon worldwide average around $8 per ton. When France attempted to impose a 0.25 cent per ton gas tax, yellow vest protests exploded—and for good reason.

In the context of a starved public sector, an across-the-board tax will hit working people the hardest; never mind the fact that the revenue generated by the gas tax was meant to fill a revenue hole left by a tax break for the wealthy. Driving a gas-guzzling vehicle to work or school isn't a choice when there's no public transit in your town or city. And after decades of wage stagnation in the US, trading in your beater for a Prius or a Tesla isn't a realistic option for many people. That fact might be hard to spot, if you only see the world through economic modeling—but it is unavoidable if you actually hope to get something passed.

Few, though, would argue that some kind of price on carbon is a bad idea. There is a compelling case to be made for including it in a much wider suite of climate policies. But market tweaks are no substitute for the robust regulation and public investment needed to build an equitable low-carbon world, with the government playing a proactive role in determining what goals the economy should be chasing.

A Green New Deal, to crib a phrase from Franklin Delano Roosevelt, should welcome the polluters' hatred. A stand-alone carbon price is scarcely more capable of solving the problem than outright denial. Given the economic transition required to avert catastrophe, Green New Dealers are right to turn to a different playbook altogether—a time in the US when large-scale economic planning didn't seem so radical. Be it the mobilization around World War II or the New Deal itself, American history is rife with examples of the state setting out a bold goal and meeting it, leveraging the full force of the US government to take on problems whose costs were simply unthinkable. There was no easy price tag to be put on human lives when elected leaders faced a depression that would leave millions starving in the streets. When the threat of the Third Reich arose, the cost of doing nothing was incalculable—and world leaders knew they would never defeat Hitler and the Axis powers by putting a modest price on fascism. Climate breakdown presents a threat no less grave or existential than those faced in the 1930s and 1940s.

From electrifying everything, to expanding mass transit, to taming the fossil fuel oligarchs—there's no path to a clean energy future that doesn't run through big government. But just one Green New Deal in one country won't be enough, either. Carbon knows no borders, and even a radically ambitious Green New Deal won't be worth much if it stops at our national borders. As it leads by example, the US will need to leverage all its might in multilateral institutions and world trade to create a global order fit for the twenty-first century, in solidarity with other countries interested in saving this increasingly balmy rock we all call home.

So where does human nature fit in all this? What if humanity is just too greedy and self-centered to make the largest economic transformation the world has ever seen—incapable of looking beyond the immediate short term to further the greater good? Not willing to sacrifice enough? The economist Rexford Tugwell, a member of FDR's so-called brain trust, might have summed it up best:

> Men are, by impulse, cooperative. They have their competitive impulses, to be sure; but these are normally subordinate. Laissez-faire exalted the competitive and maimed the cooperative impulses. It deluded men with the false notion that the sum of many petty struggles was aggregate cooperation. Men were taught

to believe that they were, paradoxically, advancing cooperation when they were defying it. That was a viciously false paradox. Of that, today, most of us are convinced and, as a consequence, the cooperative impulse is asserting itself openly and forcibly, no longer content to achieve its end obliquely and by stealth. We are openly and notoriously on the way to mutual endeavours.

THREE

MARKET FUNDAMENTALISM AT THE WORST TIME

NAOMI KLEIN

The alarm bells of the climate crisis have been ringing in our ears for years and are getting louder all the time—yet humanity has failed to change course. What is wrong with us?

Many answers to that question have been offered, ranging from the extreme difficulty of getting all the governments in the world to agree on anything, to an absence of real technological solutions, to something deep in our human nature that keeps us from acting in the face of seemingly remote threats, to—more recently—the claim that we have blown it anyway and there is no point in even trying to do much more than enjoy the scenery on the way down.

Some of these explanations are valid, but all are ultimately inadequate. Take the claim that it's just too hard for so many countries to agree on a course of action. It is hard. But many times in the past, the United Nations has helped governments come together to tackle tough cross-border challenges, from ozone depletion to nuclear proliferation. The deals produced weren't perfect, but they represented real progress. Moreover, during the same years that our governments failed to enact a tough and binding legal architecture requiring emission reductions, supposedly because cooperation was too complex, they managed to create the World Trade Organization—an intricate global system that regulates the flow of goods and services around the planet, under which the rules are clear and violations are harshly penalized.

The assertion that we have been held back by a lack of technological solutions is no more compelling. Power from renewable sources like wind and water predates the use of fossil fuels and it is becoming cheaper, more efficient, and easier to store. The past two decades have seen an explosion of ingenious zero-waste design, as well as green urban planning. Not only do we have the technical tools to get off fossil fuels, we also have no end of small pockets where these low-carbon lifestyles have been tested with tremendous success. And yet the kind of large-scale transition that would give us a collective chance of averting catastrophe eludes us.

Is it just human nature that holds us back, then? In fact, we humans have shown ourselves willing to collectively sacrifice in the face of threats many times, most famously in the embrace of rationing, victory gardens, and victory bonds during World Wars I and II. Indeed, to support fuel conservation during World War II, pleasure driving was virtually eliminated in the UK, and between 1938 and 1944, use of public transit went up by 87 percent in the US and by 95 percent in Canada. Twenty million US households—representing three-fifths of the population—were growing victory gardens in 1943, and their yields accounted for 42 percent of the fresh vegetables consumed that year. Interestingly, all of these activities together dramatically reduce carbon emissions.

Yes, the threat of war seemed immediate and concrete, but so too is the threat posed by the climate crisis that has already been a substantial contributor to massive disasters in some of the world's major cities. Still, we've gone soft since those days of wartime sacrifice, haven't we? Contemporary humans are too self-centered, too addicted to gratification, to live without the full freedom to satisfy our every whim—or so our culture tells us every day. And yet the truth is that we continue to make collective sacrifices in the name of an abstract greater good all the time. We sacrifice our pensions, our hard-won labor rights, our arts and after-school programs. We accept that we have to pay dramatically more for the destructive energy sources that power our transportation and our lives. We accept that bus and subway fares go up and up while service fails to improve or degenerates. We accept that a public university education should result in a debt that will take half a lifetime to pay off when such a thing was unheard-of a generation ago.

The past thirty years have been a steady process of getting less and less in the public sphere. This is all defended in the name of austerity, the current justification for these never-ending demands for collective sacrifice. In the past, calls for balanced budgets, greater efficiency, and faster economic growth have served the same role.

It seems to me that if humans are capable of sacrificing this much collective benefit in the name of stabilizing an economic system that makes daily life so much more expensive and precarious, then surely humans should be capable of making some important lifestyle changes in the interest of stabilizing the physical systems upon which all of life depends. Especially because many of the changes that need to be made to dramatically cut emissions would also materially improve the quality of life for the majority of people on the planet—from allowing kids in Beijing and Delhi to play outside without wearing pollution masks to creating good jobs in clean energy sectors for tens of millions.

Time is tight, to be sure. But we could commit ourselves, tomorrow, to radically cutting our fossil fuel emissions and beginning the shift to zero-carbon sources of energy based on renewable technology, with a full-blown transition underway within the decade. We have the tools to do that. And if we did, the seas would still rise and the storms would still come, but we would stand a much greater chance of preventing truly catastrophic warming. Indeed, entire nations could be saved from the waves.

So my mind keeps coming back to the question: What is wrong with us? I think the answer is far more simple than many have led us to believe: we have not done the things that are necessary to lower emissions because those things fundamentally conflict with deregulated capitalism, the reigning ideology for the entire period we have been struggling to find a way out of this crisis. We are stuck because the actions that would give us the best chance of averting catastrophe—and would benefit the vast majority—are extremely threatening to an elite minority that has a stranglehold over our economy, our political process, and most of our major media outlets.

That problem might not have been insurmountable had it presented itself at another point in our history. But it is our great collective misfortune that the scientific community made its decisive diagnosis of the climate threat at the precise moment when those elites were enjoying more unfettered political, cultural, and intellectual power than at any point since the 1920s. Indeed, governments and scientists began talking seriously about radical cuts to greenhouse gas emissions in 1988—the exact year that marked the dawn of what came to be called "globalization," with the signing of the agreement representing the world's largest bilateral trade relationship between Canada and the US, later to be expanded into the North American Free Trade Agreement (NAFTA) with the inclusion of Mexico.

The three policy pillars of this new era are familiar to us all: privatization of the public sphere, deregulation of the corporate sector, and lower corporate taxation, paid for with cuts to public spending. Much has been written about the real-world costs of these policies—the instability of financial markets, the excesses of the super-rich, and the desperation of the increasingly disposable poor, as well as the failing state of public infrastructure and services. Very little, however, has been written about how market fundamentalism has, from the very first moments, systematically sabotaged our collective response to climate change.

The core problem was that the stranglehold that market logic secured over public life in this period made the most direct and obvious climate responses seem politically heretical. How, for instance, could societies invest massively in zero-carbon public services and infrastructure at a time when the public sphere was being systematically dismantled and auctioned off? How could governments heavily regulate, tax, and penalize fossil fuel companies when all such measures were being dismissed as relics of "command and control" communism? And how could the renewable energy sector receive the supports and protections it needed to replace fossil fuels when "protectionism" had been made a dirty word?

Even more directly, the policies that so successfully freed multinational corporations from virtually all constraints also contributed significantly to the underlying cause of global warming—rising greenhouse gas emissions. The numbers are striking: In the 1990s, as the market integration project ramped up, global emissions were going up an average of 1 percent a year; by the 2000s, with "emerging markets" like China now fully integrated into the world economy, emissions growth had sped up disastrously, with the annual rate of increase reaching 3.4 percent a year for much of the decade. That rapid growth rate continues to this day, having been interrupted only briefly in 2009 by the world financial crisis. Emissions rebounded with a vengeance in 2010, which saw the largest absolute increase since the Industrial Revolution.

With hindsight, it's hard to see how it could have turned out otherwise. The twin signatures of this era have been the mass export of products across vast distances (relentlessly burning carbon all the way) and the import of a uniquely wasteful model of production, consumption, and agriculture to every corner of the world (also based on the profligate burning of fossil fuels). Put differently, the liberation of world markets, a process powered by the liberation of unprecedented amounts of fossil fuels from the earth, has dramatically sped up the same process that is liberating Arctic ice from existence.

As a result, we now find ourselves in a very difficult and slightly ironic position. Because of those decades of hard-core emitting, exactly when we were supposed to be cutting back, the things we must do to avoid catastrophic warming are no longer just in conflict with the particular strain of deregulated capitalism that triumphed in the 1980s. They are now in conflict with the fundamental imperative at the heart of our economic model: grow or die.

Once carbon has been emitted into the atmosphere, it sticks around for hundreds of years, some of it even longer, trapping heat. The effects are cumulative, growing more severe with time. And according to emissions specialists like the Tyndall Centre's Kevin Anderson (as well as countless others), so much carbon has been allowed to accumulate in the atmosphere over the past two decades that now our only hope of keeping warming below the internationally agreed-upon target of 2°C is for wealthy countries to cut their emissions by somewhere in the neighborhood of 8 to 10 percent a year. The "free" market simply cannot accomplish this task. Indeed, this level of emission reduction has happened only in the context of economic collapse or deep depressions.

What those numbers mean is that our economic system and our planetary system are now at war. Or, more accurately, our economy is at war with many forms of life on earth, including human life. What the climate needs to avoid collapse is a contraction in humanity's use of resources; what our economic model demands to avoid collapse is unfettered expansion. Only one of these sets of rules can be changed, and it's not the laws of nature.

Fortunately, it is eminently possible to transform our economy so that it is less resource-intensive, and to do it in ways that are equitable, with the most vulnerable protected and the most responsible bearing the bulk of the burden. Low-carbon sectors of our economies can be encouraged to expand and create jobs, while high-carbon sectors are encouraged to contract. The problem, however, is that this scale of economic planning and management is entirely outside the boundaries of our reigning ideology. The only kind of contraction our current system can manage is a brutal crash, in which the most vulnerable will suffer most of all.

So we are left with a stark choice: allow climate disruption to change everything about our world, or change pretty much everything about our economy to avoid that fate. But we need to be very clear: because of our decades of collective denial, no gradual, incremental options are now available to us. Gentle tweaks to the status quo stopped being a climate option when we supersized the American Dream in the 1990s and then proceeded to take it global. And it's no longer just radicals who see the need for radical change. In 2012, twenty-one past winners of the prestigious Blue Planet Prize—a group that includes James Hansen, former director of NASA's Goddard Institute for Space Studies, and Gro Harlem Brundtland, former prime minister of Norway—authored a landmark report. It stated that "in the face of an absolutely unprecedented emergency, society has no choice but to take dramatic action to avert a collapse of civilization. Either we will change our ways and build an entirely new kind of global society, or they will be changed for us."

Kevin Anderson came to a similar conclusion: "Our ongoing and collective carbon profligacy has squandered any opportunity for the 'evolutionary change' afforded by our earlier (and larger) 2°C carbon budget. Today, after two decades of bluff and lies, the remaining 2°C budget demands revolutionary change to the political and economic hegemony."

That's tough for a lot of people in important positions to accept, since it challenges something that might be even more powerful than capitalism, and that is the fetish of centrism—of reasonableness, seriousness, splitting the difference, and generally not getting overly excited about anything. This is the habit of thought that truly rules our era, far more among the liberals who concern themselves with matters of climate policy than among conservatives, many of whom simply deny the existence of the crisis. Climate change presents a profound challenge to this cautious centrism because half measures won't cut it: an "all of the above energy" program, as US president Barack Obama described his approach during his two terms in office, has about as much chance of success as an all-of-the-above diet, and the firm deadlines imposed by science require that we get very worked up indeed.

The challenge, then, is not simply that we need to spend a lot of money and change a lot of policies; it's that we need to think differently, radically differently, for those changes to be remotely possible. A worldview will need to rise to the fore that sees nature, other nations, and our own neighbors not as adversaries, but rather as partners in a grand project of mutual reinvention.

That's a big ask. But it gets bigger. Because of our endless procrastination, we also have to pull off this massive transformation without delay. The International Energy Agency (IEA) warns that "we have no room to build anything that emits CO_2 emissions," if we are to have any hope of keeping warming below 2 degrees, never mind the 1.5-degree threshold that vulnerable nations have demanded. Fatih Birol, the executive director of the IEA, warned that the entirety of our carbon budget—the fossil fuels we can burn while staying below 2 degrees—will be eaten up by existing fossil fuel infrastructure. All new major infrastructure projects need to be low-carbon, or existing fossil fuel infrastructure, including power plants, buildings, and cars, need to be replaced earlier than planned.

All this means that the usual free market assurances—A quick techno-fix is around the corner! Dirty development is just a phase on the way to a clean environment, look at nineteenth-century London!—simply don't add up. We don't have a century to spare for China and India to move past their Dickensian phases. Because of our lost decades, it is time to turn this around now. Is it possible? Absolutely. Is it possible without challenging the fundamental logic of deregulated capitalism? Not a chance.

While doing this research, I was struck by a mea culpa of sorts, written by Gary Stix, a senior editor of *Scientific American*. Back in 2006, he edited a special issue on responses to climate change, and, like most such efforts, the articles were narrowly focused on showcasing exciting low-carbon technologies.

But in 2012, Stix wrote that he had overlooked a much larger and more important part of the story—the need to create the social and political context in which these technological shifts stand a chance of displacing the all too profitable status quo. "If we are ever to cope with climate change in any fundamental way, radical solutions on the social side are where we must focus, though. The relative efficiency of the next generation of solar cells is trivial by comparison."

In other words, our problem has a lot less to do with the mechanics of solar power than the politics of human power—specifically whether there can be a shift in who wields it, a shift away from corporations and toward communities, which in turn depends on whether or not the great many people who are getting a rotten deal under our current system can build a determined and diverse enough social force to change the balance of power. Such a shift would require rethinking the very nature of humanity's power—our right to extract ever more without facing consequences, our capacity to bend complex natural systems to our will. This is a shift that challenges not only capitalism but also the building blocks of materialism that preceded modern capitalism, a mentality some call "extractivism."

Because, underneath all of this is the real truth we have been avoiding: climate change isn't an "issue" to add to the list of things to worry about, next to health care and taxes. It is a civilizational wake-up call. A powerful message—spoken in the language of fires, floods, droughts, and extinctions—telling us that we need an entirely new economic model and a new way of sharing this planet. Telling us that we need to evolve.

Some say there is no time for this transformation; the crisis is too pressing and the clock is ticking. I agree that it would be reckless to claim that the only solution to this crisis is to revolutionize our economy and revamp our worldview from the bottom up—and anything short of that is not worth doing. There are all kinds of measures that would lower emissions substantively that could and should be done right now. But we aren't taking those measures, are we? The reason is that by failing to fight these big battles that stand to shift our ideological direction and change the balance of who holds power in our societies, a context has been slowly created in which any muscular response to climate change seems politically impossible.

On the other hand, if we can shift the cultural context even a little, then there will be some breathing room for those sensible reformist policies that will at least get the atmospheric carbon numbers moving in the right direction. And winning is contagious. So who knows?

For a quarter of a century, we have tried the approach of polite incremental change, attempting to bend the physical needs of the planet to our economic model's need for constant growth and new profit-making opportunities. The results have been disastrous, leaving us all in a great deal more danger than when the experiment began.

Climate change is already here, and increasingly brutal disasters are headed our way no matter what we do. But it's not too late to avert the worst, and there is still time to change ourselves so that we are far less brutal to one another when those disasters strike. And that, it seems to me, is worth a great deal. The thing about a crisis this big, this all-encompassing, is that it changes everything. It changes what we can do, what we can hope for, what we can demand from ourselves and our leaders. It means there is a whole lot of stuff that we have been told is inevitable that simply cannot stand. And it means that a whole lot of stuff we have been told is impossible has to start happening right away.

Can we pull it off? All I know is that nothing is inevitable. Nothing except that climate change changes everything. And for a very brief time, the nature of that change is still up to us.

FOUR

AVERTING CLIMATE COLLAPSE REQUIRES CONFRONTING RACISM

IAN HANEY LÓPEZ

The Green New Deal promises aggressive measures to avert climate collapse. In addition, though, it also proposes radical measures to ameliorate economic inequality and racism. The Sunrise Movement describes it as "the only plan put forward to address the interwoven crises of climate catastrophe, economic inequality, and racism at the scale that science and justice demand." Yet even a large number of progressives see dangerous overreach in the legislation's linking of the environmental emergency with the fights for economic as well as racial justice. Many worry that seeking to simultaneously address the environmental crisis, capitalism, and racism thrills activists but virtually guarantees that nothing happens to save the planet. On the contrary: enacting the Green New Deal *requires* addressing race, class, and climate together.

To see why, step back for a moment to consider the animating insight behind the Green New Deal. To slow global warming, the United States government must intervene on a massive scale. No salvation is possible without government's might. What stands in the way? The primary culprit seems to be the reigning free market ideology that slanders government as serfdom's handmaiden and instead heralds loosely regulated capitalism as the surest route to liberty. This ideology prevents the government from acting forcefully on behalf of most Americans, thereby condemning the vast majority to chronic economic, health, and environmental jeopardy while it concentrates wealth in corporations and family dynasties. Yet it continues to hold sway not just in the halls of power but among broad swaths of voters.

So the question repeats itself, now pushed back one level. Preventing climate collapse depends upon the government acting forcefully. Government action requires defeating free market ideology. Again, we must ask: What stands in the way?

Racism—or, more particularly, the Right's strategic manipulation of racial resentment. The intense concentration of society's wealth generates widespread social misery. What can justify awarding the ultrarich so much while most struggle to get by on so little? Stories about free markets are not nearly sufficient to this task. They are far too abstract. Instead, the Right deploys visceral racist narratives about who "we" are, who threatens us, and who protects us, all told in coded language to hide the racism while nevertheless triggering racial resentment and breaking public confidence in government for the collective good.

Decades of right-wing narratives linking people of color, hostility toward government, and class war have culminated in a strong connection between racial resentment and climate denial. "Democratic voters are almost twice as likely as Republicans to agree with the scientific consensus on climate change," the Sierra Club reported in 2018. This partisan divide is deeply contoured by racial animosity: "white Republicans who scored at the highest level for racial resentment were over three times as likely to disagree with the statement that climate change was real than white Republicans who tested at the lowest level of the scale." Among the Republicans measuring the highest in racial resentment, more than four out of five, 84 percent, denied that climate change is caused by humans.

Many on the broad left now see clearly that racial terms permeate the Right's political rhetoric. Think of coded phrases like "urban disorder," "thugs," and "welfare queens." Or terms like "the silent majority," "real Americans," and "the heartland." Or "illegal aliens" and "make America great again." This is dog whistling, the spreading of powerful racial narratives expressed in terms that on the surface do not mention race.

Less evident to many progressives is the role these poisonous words play in shifting attitudes toward government and thereby facilitating rule by the rich. Since the civil rights movement in the 1960s, however, dog whistle politics has been the Republican Party's principal route to power. Formerly widely viewed as the party of big business, the GOP has remade its public image, recasting itself as the defender of the (white) working class and (white) evangelical Christians—even as it continues to govern in the interests of its dark money patrons. Donald Trump epitomizes this politics, strategically exploiting coded racism to build popular support for government by and for billionaires. But of course it's not just Trump. It's the entire GOP (plus some Democrats), who serve plutocrats rather than the people. For these politicians, obeisance to the wealthy includes blocking climate action that would reduce the profits of the world's largest polluters.

So here's where we are: to fully mobilize government and avert climate collapse will require a wave election that breaks the political stalemate of recent decades, not only taking the presidency but also putting progressives in charge of the House and Senate. Yet racial resentment gives the Republican Party staying power—and, relatedly, racism fuels opposition to government efforts to avert climate collapse. In short, averting climate collapse requires massive government action, which depends on a wave election that breaks the GOP's death grip on political power, which requires confronting racism as a weapon in the class war the rich are winning.

The prospect may seem daunting—a "Today's to-do list" that starts with "end global warming," "confront racism," and "remake capitalism," before listing things needed from the grocery store. But progressives already have a plan for defeating the three horsemen of doom riding together—global warming, capitalism, and racism. It's called the Green New Deal. Good research shows that simultaneously addressing class, race, and government is not just possible; it's actually the best way to build a multiracial progressive wave. This chapter concludes on these upbeat notes. But first, the bulk of the chapter exposes the deep symbiosis between plutocracy and racism in the United States.

RACE, CLASS, AND GOVERNMENT

Understanding the preconditions for the next New Deal that tackles climate change calls for understanding how racism has been weaponized in American politics. That's because dog whistle politics played an outsize role in destroying the original New Deal.

Historians refer to the "long New Deal" to describe the decades spanning the 1930s through the 1960s. Across these four decades, a coalition primarily comprised of African Americans, the white working class, and coastal liberals often elected Democratic politicians. These politicians—and eventually most Republican elected officials as well—believed that government should regulate the marketplace, build a safety net, support unions, invest in infrastructure, and use high taxes to redistribute wealth downward. To be sure, government programs during these decades systematically discriminated against non-whites and women. Still, Democrats undertook actions that improved their economic and social position, making women and African Americans indispensable members of the New Deal coalition. The Democratic Party pursued both economic liberalism and civil rights, realizing both were essential to winning elections.

But over those same decades, the New Deal's very successes created the conditions for a perverse backlash. The New Deal pulled millions of whites out of the destitution deepened by the Great Depression. In turn, many new members of the middle class came to imagine themselves masters of their own destiny rather than beneficiaries of government concern. Meanwhile, civil rights triumphs that upended the racial order prompted many whites to increasingly oppose meaningful racial integration in workplaces, schools, and neighborhoods.

Sensing opportunity especially in the white racial resentment developing in response to the civil rights movement, a reactionary faction within the GOP formulated a plan to shatter the New Deal coalition. The conservative journalist Robert Novak was a fly on the wall when that plan coalesced during the 1963 meeting of the Republican National Committee in Denver. Novak laid that new plan bare, reporting that "a good many, perhaps a majority of the [Republican] party's leadership, envision substantial political gold to be mined in the racial crisis by becoming in fact, though not in name, the White Man's Party."

"The racial crisis"? We call that the civil rights movement.

"In fact, though not in name"? The civil rights movement had made open endorsements of white supremacy political suicide. The GOP would use code.

"The White Man's Party"? Yes, but as strategy, not bigotry. Up to that point, it was the Southern Democrats who had fashioned themselves as the White Man's Party, using fraud and violence to ensure that virtually no blacks voted. Now the GOP, up to then relatively racially moderate, would transmogrify. Richard Nixon called this his "Southern strategy." As early as 1972, though, it became clear that coded racial appeals aimed at whites would work across the country.

This racial strategy was not, however, exclusively about race. Listen to the Republican operative Lee Atwater. With roots in South Carolina politics, Atwater held Nixon as a personal hero, describing his race-baiting as "a blueprint for everything I've done in the South." In 1981, Atwater gave what was then an anonymous interview. He was a rising star, on his way to becoming the political director of Ronald Reagan's 1984 reelection campaign, the manager of George Bush's 1988 presidential campaign, and eventually the chair of the Republican National Committee. Perhaps lulled by the promised anonymity of the interview, Atwater candidly explained the Right's use of coded racism to win elections:

> You start out in 1954 by saying, "Nigger, nigger, nigger." By 1968 you can't say "nigger"—that hurts you. Backfires. So you say stuff like forced busing, states' rights, and all that stuff, and you're getting so abstract now, you're talking about cutting taxes, and all these things you're talking about are totally economic things and a byproduct of them is, blacks get hurt worse than whites . . . Obviously sitting around saying, "We want to cut taxes and we want to cut this," is much more abstract than even the busing thing, and a hell of a lot more abstract than "Nigger, nigger." So any way you look at it, race is coming on the back burner.

From "nigger, nigger, nigger" to "states' rights" and "forced busing," and from there to "cutting taxes," always with race "coming on the back burner."

"States' rights" pretended to refer to federal respect for state sovereignty, though as a dog whistle in the 1950s and 1960s, it commonly meant the right of southern states to use laws and violence to oppress and humiliate African Americans. "Forced busing" provided a superficially race-neutral means to criticize the public school integration that busing was designed to achieve. What about tax cuts? "Blacks get hurt worse than whites," explained Atwater.

Atwater's direct linkage of anti-black racism to tax cuts opens a new vista on dog whistle politics. The Right uses racial appeals to sunder the political alliance of working-class whites with African Americans and coastal liberals. But it also uses racism to do much more than shatter the New Deal coalition. It exploits racism to break New Deal commitments—the belief that government should actively work for people rather than for concentrated wealth. It does so by reframing activist government as primarily concerned with helping African Americans. Opposing strong government becomes, supposedly, a way to restrict help to blacks. In reality, of course, this curtails government's ability to work for working families of every color.

The race-class-and-government narratives used by the Right over the last half century can be stripped down to their bare bones. Doing so sacrifices nuance, but it illuminates how the Right has exploited racism to shift popular attitudes toward government in a manner that helps the very wealthiest. These are the narrative bones exploited by the Right to march its class agenda forward: 1. Fear and resent people of color; 2. Distrust government; 3. Trust the marketplace. The Right tells other stories as well. Some involve additional group resentments—for instance, patriarchy and homophobia. Others are more purely economic; for example, the constant invocations of the free market. Still, separating out the core elements of the Right's *racial* story helps make clear today's deep connections between racism, democracy, and plutocracy.

1. *Fear people of color because their basic nature is violent; resent them because they are lazy and rip off government rather than do honest work.*

   The core narrative is highly racialized, a basic story about whites threatened by people of color. But it is not told in those naked terms. Instead, the Right translates this core narrative into dog whistles that convey a more diffuse conflict between innocents and criminals, welfare queens versus hard workers. This is key because of how it shrouds the racism from whites. It's also critical, though, because these less starkly racial terms help sell the Right's ideas even to many people of color. In its most naked form, few people of color and indeed relatively few whites would accept a bald story about white virtue and minority venality. Dog whistling hides the essential racism and transforms this core narrative into "common sense" accepted by many who repudiate racism, including many people of color.

2. *Distrust government because it doesn't care about white people. Instead, it coddles undeserving minorities with welfare spending and it refuses to control violent minorities by leaving laws unenforced and borders open.*

   There are numerous reasons beyond race for skepticism toward government. This includes government as a swamp of self-dealing by special interests and corrupt politicians. For communities of color, government overpolicing and mistreatment also stoke distrust. Nevertheless, the Right relies especially heavily on racial resentment to fuel hostility toward government. The right-wing echo chamber booms with messages warning that liberal government, rather than protecting whites, instead sacrifices them in order to pamper lazy people of color. Government, the Right insists, is the enemy of (white) working people.

3. *Trust the marketplace because the market rewards hard work. Support the very wealthy and large corporations as the job creators.*

   This is the least convincing element to many voters. Right-leaning voters have high confidence that hard work is rewarded, but more mixed feelings toward the wealthy, with many seeing them as rigging the rules. Trust in the marketplace gains credibility often as a default position, the only alternative once one loses confidence in collective action and government. At its heart, the Right promotes the notion that everyone is on their own and must fend for themselves and their families. On the level of economic security, it means trusting the marketplace. On the level of physical safety, it often means buying a gun. For many, this is no more than an act of desperation once they've come to believe that government cares about others, not them.

Many progressives find it bewildering that most white working people hate government and trust the rich (or, in fancier language, they're confounded by a populace that embraces neoliberalism). This is especially dumbfounding from a historical point of view that looks back to the New Deal. In that era, the great masses of Americans understood that their salvation depended on collective action through unions as well as government. What happened? This question cannot be answered from a vantage point that looks only at the rise of pro-market, shrink-the-government ideology. On the contrary, it's only when we recognize how the Right manipulates racial resentment that we can grasp what happened to the vast American middle class: all too many of them voted their racial fears, thereby betraying not just racial justice but their own economic futures. The same dynamic continues to play out today, though now, the chance to slow climate change is also at stake.

RACISM, GOVERNMENT, AND GLOBAL WARMING

The immense fortune of Charles Koch rests on a huge petrochemical and industrial conglomerate. For decades, he and his brother David spent barrels of cash protecting their bottom line against efforts to protect the environment. Initially, they focused on sowing doubt about climate science and buying cooperation from pliant politicians. It was devastatingly effective from behind the scenes. As Kert Davies, the director of the Climate Investigations Center, explained to the journalist Jane Mayer in 2019, "You'd have a carbon tax, or something better, today, if not for the Kochs. They stopped anything from happening back when there was still time."

By 2009, though, funding climate denial and purchasing loyal allies in Congress was not enough. Newly elected president Barack Obama promised action on the environment, including regulations to raise the cost of emitting the greenhouse gases warming the world's atmosphere. Mayer is the author of *Dark Money*, a book carefully scrutinizing the Koch brothers and the circle of wealthy donors they enlisted to fight for rule by the rich. She summarized the views held by the Koch brothers at the outset of the Obama administration: "Koch Industries regarded any compromise that might reduce fossil fuel consumption as unacceptable. Protecting its fossil fuel profits was, and remains, the company's top political priority." She adds that "the Kochs, to achieve this end, worked to hijack the Tea Party movement and, eventually, the Republican Party itself."

The Tea Party? This movement exploded on the scene in the first years of the Obama administration. Some of its early energy came from genuine economic grievances generated by a severe recession. Very quickly, however, the Right harnessed this seething fury, shifting the blame for economic misery from the plutocrats to people of color. One critic summed up the Tea Party this way: "It's a revolt against the very notion of a positive role for government in helping people. It's a revolt against Latin American immigrants. It's a revolt against Muslim Americans. And it's a revolt against our black President." Excessive government, immigrants, Muslims, and a black president: the perfect storm of dog whistle themes.

If racial resentment howled through the Tea Party, the funding that grew this grassroots rage into a political behemoth came from big-money donors, chief among them the Koch brothers. According to a Republican campaign consultant, the Koch brothers "gave the money that founded [the Tea Party]. It's like they put the seeds in the ground. Then the rainstorm comes, and the frogs come out of the mud—and they're our candidates!" As Mayer concluded at the time, "the anti-government fervor infusing the 2010 elections represents a political triumph for the Kochs. By giving money to 'educate,' fund, and organize Tea Party protesters, they have helped turn their private agenda into a mass movement." Or as David Axelrod, a senior adviser to Obama, acidly remarked, "This is a grassroots citizens' movement brought to you by a bunch of oil billionaires." Bottom line: when the Kochs contrived to protect their petrochemical empire and to stop climate legislation, their main tactic was to fund racial division.

The sociologist Arlie Hochschild spent several years during the Obama era interviewing Tea Party activists in Louisiana. Hochschild wanted to understand why white Louisianans so fiercely hated environmental regulations, when they so desperately needed these laws to protect the lands and waters on which their families' livelihoods depended. To explain this conundrum, she ultimately settled on a metaphor of people cutting in line:

> You are patiently standing in the middle of a long line stretching toward the horizon, where the American Dream awaits. But as

> you wait, you see people cutting in line ahead of you. Many of these line-cutters are black—beneficiaries of affirmative action or welfare. Some are career-driven women pushing into jobs they never had before. Then you see immigrants, Mexicans, Somalis, the Syrian refugees yet to come. As you wait in this unmoving line, you're being asked to feel sorry for them all. You have a good heart. But who is deciding who you should feel compassion for? Then you see President Barack Hussein Obama waving the line-cutters forward. He's on their side. In fact, isn't he a line-cutter too? How did this fatherless black guy pay for Harvard? As you wait your turn, Obama is using the money in your pocket to help the line-cutters. He and his liberal backers have removed the shame from taking. The government has become an instrument for redistributing your money to the undeserving. It's not your government anymore; it's theirs.

Though Hochschild does not frame it this way, her narrative records the essential teachings of dog whistle politics. Racial resentment, government betrayal, reliance on the free market, these were the essential teachings of the Right.

They are also the core campaign themes of Donald Trump. His election has been a gusher for extractive industries, including those of the Koch brothers. Jane Mayer reports: "Whether announcing his intention to withdraw from the Paris climate accord, placing shills from the oil and coal industries at the head of federal energy and environmental departments, or slashing taxes on corporations and the ultra-wealthy, Trump has delivered for the Kochs." Charles Koch himself boasts these are the golden years for his agenda. As he told his allied political donors in 2018, "We've made more progress in the last five years than I had in the previous fifty." The story of the widespread hostility toward averting climate collapse is the story of dog whistling in microcosm—though now the fate of the world hangs in the balance.

CONCLUSION: THE GREEN NEW DEAL

The basic strategy of the Right—including major polluters like the Koch industries—is to fund racist dog whistling as a means to break popular confidence in government. In turn, creating a progressive wave will require mobilizing a broad movement that rejects racial division in favor of cross-racial solidarity. To be crystal clear, this movement must tackle racism. It cannot succeed simply by focusing on concerns like the economy or the environment alone.

After the 2016 election, I co-founded a major research project on how to beat dog whistle politics. Using focus groups and extensive polling, we set out to test the power of various progressive messages. Our striking, encouraging findings became the seed of my book, *Merge Left: Fusing Race and Class, Winning Elections, and Saving America*. With one survey, we tested economic populism. In a nationally representative poll of 2,000 voters, we asked respondents which came closer to their views: a message promoting the nostrums of deregulated capitalism or one promoting a progressive economic agenda:

> "To make life better for working people we need to cut taxes, reduce regulations, and get government out of the way."
>
> or
>
> "To make life better for working people we need to invest in education, create better-paying jobs, and make health care more affordable for people struggling to make ends meet."

The progressive message beat the unrestrained-capitalism message by a whopping 32 points. This result confirms that voters generally prefer progressive over pro-business economic policies.

And yet the Right keeps winning. How? By not campaigning on economic messages alone, or even primarily. Instead, the Right exploits coded racial stories about "terrorist countries," "illegal immigration," and "criminal gangs" that threaten "hardworking Americans," "our communities," and "our own people."

In our polling, we also tested a message of racial fear. Against it, the advantage of the progressive economic proposals evaporated. Overall, voters rated the progressive economic message as being a bit less convincing than the Right's racist insinuations. In other words: if the Left avoids tough conversations about racial division, it might win some elections, but the country will remain locked in a political stalemate that guarantees that no big government efforts to save the planet will ever be enacted.

To respond to dog whistle divisions, we developed a countervailing core narrative urging people to (1) distrust greedy elites sowing division, (2) join together across racial lines, and (3) demand that government work for all racial groups, whites included. These elements provided the scaffolding on which we built nine different messages. The race-class story they told proved remarkably powerful.

True, the roughly one in five voters who form the reactionary base hated the race-class messages. But progressives embraced them. Even more importantly, persuadable voters, the almost six in ten in the swing middle, strongly preferred our messages of cross-racial unity to the Right's narrative promoting racial resentment. Persuadable voters found *all nine* of our messages more convincing than the dog whistle racial fear message. These were the same voters who favored the racial fear message over economic populism. Bottom line: it's not just possible to talk about race, class, and government together. It's the most potent political message circulating today, right or left.

When Green New Dealers say that the planet-destroying schemes of fossil fuel millionaires depend on them successfully dividing us against each other, this young movement showcases the power of framing racism as a weapon of the rich. Indeed, the Green New Deal provides a vehicle to help build a progressive multiracial wave.

Rather than only viewing the Green New Deal as a piece of proposed legislation, it makes more sense to see it also as a vehicle for building political power. The goal of averting climate catastrophe is immediate and galvanizing. But in a political world warped by racist narratives, even an urgent goal is not enough. Instead, the fight to save the environment must be contextualized in terms of a compelling story about who we are, who threatens us, and how we move forward. It must, in other words, invoke a narrative of cross-racial solidarity—only together can we defeat the economic titans who intentionally stoke division so we can build a better world for all our families. This is the genius of the Green New Deal. It lays the groundwork for connecting environmental activism to the fights for racial and economic justice.

Today, some of society's most powerful elements seek to divide us, and their success exposes us to climate death. Mobilizing in unison to confront these threats is the only way to build and nurse the massive collective response necessary to respond to climate catastrophe. To succeed, that same mobilization must challenge plutocracy as well as racial hierarchy. Averting climate collapse is a concrete policy goal that can help build a multiracial, class-conscious coalition that will make other progressive dreams possible, too, from ending mass incarceration and mass deportation to providing affordable and excellent education and health care. Collective empowerment is the precondition for creating a society in which all of our families can thrive. The urgency of the fight to save the planet can impel us to save our democracy and society as well.

PART II

GREEN NEW DEAL VISIONS AND POLICIES

FIVE

HOW WE GOT TO THE GREEN NEW DEAL

BILL McKIBBEN

The most commonly asked questions about the Green New Deal may be: "Why's it so big? Why does it move so fast?" And those are good questions: it is a sizable and sprawling remake of much of America's domestic policy, in some ways the biggest set of policy proposals since the namesake New Deal of the Depression.

So let me provide some history that helps explain why we find ourselves in this spot. I'm probably in as good a position as anyone else to supply it, because I've lived through most of the struggle for climate action since the day in 1988 when NASA scientist James Hansen opened the public phase of the fight over what we then called the greenhouse effect—with his historic testimony before Congress. A year later, in 1989, I published what is generally regarded as the first book for a general audience on the topic, *The End of Nature.* What we understood then was unambiguous: First, when you burned coal and gas and oil, you poured carbon dioxide into the atmosphere. And second, the molecular structure of carbon dioxide trapped heat.

At first the reaction was what one would expect. Having been warned of the most dangerous problem the world has ever faced, political leaders reacted appropriately, pouring money into research to make sure we understood the dimensions of our peril and pledging to take action. George H. W. Bush, the Republican president at the time, promised that we would fight "the greenhouse effect with the White House effect," which was a pretty good line.

Had we taken action then—and this is important to realize—we had many fairly modest steps that would have yielded big results. For instance: economists called for a tax on the greenhouse gas emissions—primarily CO_2 and methane—that cause global warming. The point of that tax would have been to send a signal: we need to start scrubbing fossil fuel out of our economy because it's dangerous. And had we enacted some manageable levy at that point, the results would have played out over time. The supertanker that is our global economy would have begun a slow and ponderous turn, measured at first in just a few degrees of direction. But if you start steering left, over time that course correction lets you sail into an entirely different sea. I'm not sure we would have *solved* global warming thirty years later, but we'd be well on the way.

We didn't do that, however. Instead we did the opposite: we began burning fossil fuels ever more quickly. We not only kept the supertanker heading in precisely the same direction, we shoveled on the coals (literally—this is only sort of a metaphor). Human beings have produced more greenhouse gases in the three decades since Jim Hansen's testimony than in all of human history before that time.

And so in the course of those decades we've gone from a potentially dangerous but perhaps still manageable problem to a crisis that requires truly dramatic intervention if we're to have any hope. In the late 1980s, reducing carbon emissions 1 or 2 percent a year would have sufficed; now that we're actually at the cliff's edge, we have to move far far more urgently—5, 6, 7 percent a year, which is on the outer edge of the possible. It will take a society-wide effort, of precisely the kind envisioned by the Green New Deal.

So what happened?

We know the answer to this a lot more clearly than we did at the time, when there was only a general sense that *something* was slowing down our response. Thanks to intrepid reporting from the likes of *InsideClimate News*, the *Los Angeles Times*, and the Columbia Journalism School, we understand better what happened. As it turned out, the big fossil fuel companies had known well before Hansen what was happening. They'd begun a serious study of global warming in the 1970s, as supercomputers began to become fast enough to model the climate. Exxon, for instance, was in those days the biggest company on earth, with a crack staff of scientists, and its *product was carbon.* Of course it was going to get to the bottom of it. The company outfitted a tanker with CO_2 monitors; its scientists produced accurate graphs of how fast CO_2 concentration and temperature would rise. (Their predictions have since been proven startlingly accurate, with carbon levels today basically in line with their graphs.) And they were believed by their bosses: Exxon began building its drilling rigs higher to compensate for the rise in sea level they knew was on the way.

But what they didn't do was tell the rest of us. Instead, the oil companies and the coal companies and the utilities, with some help from the rest of the fossil-fuel-intensive industry, began to studiously build the architecture of deceit and denial and disinformation that has kept us ever since locked in an utterly phony debate about whether or not global warming is "real"—a debate, remember, that both sides have known the answer to from the start. It's just that one side was willing to lie.

That lie began to take shape during the 1990s, as industry forums like the Global Climate Coalition began to pump out disinformation. They'd hired many veterans of the tobacco wars, and hit on many of the same strategies. There was, they claimed, serious doubt within the scientific community (and anyway it was people's fault for burning gas, just as it had been their fault for smoking). But the momentum to tackle this problem was strong: people had been seriously scared by the early warnings. And so the world began to move toward Kyoto, Japan, and the first attempt, at the end of the 1990s, to actually set some new rules.

Before they could get there, the oil industry hit hard: the most famous speech came from Lee Raymond, then head of Exxon, who told a gathering in Beijing, in the run-up to Kyoto, not to worry—that the planet was cooling. That lie was coupled with others: that climate action would sabotage economic growth in the developing world, and that "it is highly unlikely that the temperature in the middle of the next century will be significantly affected whether policies are enacted now or 20 years from now." That is precisely the opposite of the truth—we knew then and now that early action would make all the difference—but it was enough. I was at the Kyoto talks, and I'll never forget them. The world did reach a last-minute agreement, but as everyone applauded and cheered I was standing next to the lobbyist, since deceased, who'd been coordinating the opposition. "I can't wait to get back to Washington, where we've got this under control," he said. I thought he was whistling past the graveyard. But he was right.

The next decade—the 2000s, with George Bush the younger at the helm most of the way—saw the unchallenged power of the fossil fuel industry. Year after year they set records for new profits, and successfully prevented any action on climate change. (It was just days into Bush's presidency that Lee Raymond came to the White House for a visit with his old friend Dick Cheney, who had been running oil service giant Halliburton before becoming vice president, and soon after that the administration dropped its promise to treat CO_2

as a pollutant.) Public concern grew—it was also the decade of Al Gore's *An Inconvenient Truth*—but when the world met in Copenhagen near the end of the decade to try to reach a global treaty, it went badly indeed—despite remarkable hype, essentially nothing happened. And shortly after that, when Congress tried to debate the Waxman-Markey bill calling for caps on carbon emissions, it was a rout: the Senate didn't even bother to hold a vote because the outcome was so clear. The industry had carried the day—there would be no action on climate.

In the Obama years, we struck back in force, and gained a little ground, though not nearly enough. We built a movement that relentlessly fought the Keystone XL pipeline for nearly a decade, until the White House finally decided to reject TransCanada's construction permit. We divested trillions of dollars from the fossil fuel industry, by engaging people across the world to pressure their university, or their town council, to pull all their money out of companies profiting from the wrecking of our planet. And we shocked the conscience of the nation during the fight over the Dakota Access Pipeline, when Americans saw vivid images of police officers hosing down Native American protesters in the frigid Dakota winter, as they silently insisted that water is life. In each of these fights our movement raised another inconvenient truth into the national debate: the fossil fuel industry's business plan is incompatible with a livable future, and their business depends on poisoning the water, air, and land of people with little power to resist.

Those fights were all meaningful—Alexandria Ocasio-Cortez, now a visionary leader in the fight against the climate crisis, credits the Dakota Access fight with her radicalization on the issue. And they resulted in enough pressure to force the Obama administration to enact a Clean Power Plan and a ban on coal extraction on public lands. But we were losing the larger war, and badly. Temperatures kept soaring, even as the White House lectured climate activists about why one pipeline wasn't worth fighting over.

It's important, though, to understand something else, which is that the spike in temperature was not happening in a vacuum. Something else was spiking, too, and that was the level of inequality in our economy.

Remember that in the years after the original New Deal, inequality in our country was declining. Incomes rose steadily for the bottom 90 percent of the population, but high marginal tax rates meant that the wealth of the 1 percenters was stagnating. The year I graduated from high school, 1978, was also the year the top 1 percent of Americans saw their share of the nation's wealth fall to 23 percent, the lowest it has ever been; we seemed to be on course for an ever-fairer society.

And then we had the Reagan revolution, and all that changed. Many members of his cabinet described the extremist libertarian Ayn Rand as their favorite philosopher; her acolyte, Alan Greenspan, rose to planet-spanning dominance as the head of the Federal Reserve; and the cult of laissez-faire, markets-solve-all-problems economics took full control of our system. Tax rates dropped, and the wealth gap began to open wider. After thirty-five years of this, capped by the Trump tax cuts of 2018, it's no wonder that we've reached a level of gross, cartoonish inequity: the eight richest men in the world have more wealth than the bottom 3.6 billion people. The six heirs to the Walmart fortune have more money than the bottom 40 percent of the American population. Half of our fellow Americans don't have the few hundred dollars in reserve to get them through an emergency. Medical bankruptcies plague the country; student debt continues to spiral. People live in the very real fear that if they lose their jobs the rest of their lives will crumble, since the holes in the social safety net are now so large that it's downright scary to walk the tightrope that life requires.

The rise of inequality was not unconnected to the rise of the temperature. The same people were working hard to deny both climate science and social ethics: one need look no further than the Koch brothers, who were simultaneously the biggest oil and gas barons in the country and the most influential political players. They used their cash to deploy an endless series of front groups around the country, many of which were devoted to blocking renewable energy, defunding mass transit, and trashing environmentalists. At the same time, they were working hard to lower tax rates for the rich, depress social spending, and make life ever harder for the mass of people at the bottom of the economic pile.

Again—you can't separate these phenomena, and not just because the same people were pushing both disasters. It's because of their quest to consolidate ever more wealth and power in the hands of the few that working people's lives have gotten harder over the past few decades, and why the climate system has been allowed to be brought to the point of collapse. It's also because one effect of the increased inequality was to make people ever less secure, and so ever more fearful of change.

What that means, I think, is that if we want to address one of the two great disasters of our time—the spiking temperature—then we have to address the other as well—the spiking levels of inequality. They both loom so large that they can't be taken on piecemeal: if you accept the argument that we need dramatic change to deal with the climate crisis, then there's no way to get change on that scale without simultaneously addressing the economic insecurity that makes it so hard for people to imagine change. We need, in other words, to do just what the Koch brothers pulled off, but in reverse.

If we try to do one without the other, then the politics become impossible. Look at what happened in France, when president Emmanuel Macron tried to raise gas taxes to help meet the country's carbon emissions targets: seemingly out of nowhere the yellow vest movement arose, among people—often rural and poor—for whom those taxes were an insupportable burden. That's precisely the kind of message the fossil fuel industry will do everything it can to spread in this country: for one example, look at the vast floods of money they poured into referendums on a carbon price in the state of Washington, or even a Colorado proposal to move fracking wells out of people's backyards and school zones. Always the message was the same: this will cost you money. And it's a message that resonates with people who don't have money. Keeping someone insecure is a perfect way to make sure they're scared of change, because if things got any worse, then life would be impossible. The painful truth is, things can always get worse, and no one knows that better than people for whom things are already bad.

And hence the political shrewdness of the Green New Deal, with its recognition that we're either going to deal with these problems at the same time, or not at all. The energy transformation that physics absolutely requires, requires in turn that we mobilize our societies in ways we haven't seen since the lead-up to World War II (not surprisingly, the same years when inequality began to narrow). And that's why, for instance, a federal job guarantee is such good public policy: it not only provides the person the power required to put up the solar panels and insulate the walls but also provides the basic underlying insurance that societies require if they're going to take chances. And we are going to have to take chances.

The good news—and it's very good news—is that a movement has risen in response to these crises, a movement rooted in concerns for both science and justice. And it's a movement that's learned some lessons, all of them incorporated in the Green New Deal. For one, the fossil fuel industry is always going to fight. And two, the best response is a big, broad movement. The people most passionate about solving the climate crisis, the ones shutting down coal plants and pipelines, are often those who have contributed little to the problem and also struggle against other injustices—immigrants facing climate crisis at home and draconian immigration policy, African Americans fighting polluting coal plants and economic disinvestment, young people looking out on piles of student debt and an uninhabitable future. That is to say, they are ordinary people. And for ordinary people, the climate is not an issue that can or should be isolated from the economy, or national identity, or justice. Some advocates in DC had long thought that the key was to simplify the issue, when the truth is that the base of our movement already lives at the intersections of this crisis. It's good movement-building and good politics to put forward a vision that confronts the crises of inequality, injustice, and climate breakdown.

So many of the best leaders of the climate movement aren't people you've necessarily heard of: scientists, professors, or Hollywood actors (though they, too, have places of honor—Sandra Steingraber, Naomi Oreskes, Mark Ruffalo; there's an endless list like them). Instead, they're community leaders in the places that have been hardest-hit by the fossil fuel industry and the rising temperature. Sometimes that's overseas, where the damage is the steepest. I've been to Bangladesh and watched people die of dengue fever; I've stood on the Greenland ice shelf with a poet from the Marshall Islands as she raged about the fact that the snow melting beneath her feet would soon drown her homeland. Just as often, it's close to home. I've been arrested at the gates of the Chevron refinery in Richmond, California—a plant so noxious that every local resident has to monitor texts from the company in case of a release of toxic gases. I've watched the incredible bravery of the nonviolent protesters at

Standing Rock in the Dakotas, as they tried to face down a pipeline that threatened their water.

All of this taught me things I needed to know about the way climate change and fossil fuel intersect with other crises. People are short of money in part because they pay huge proportions of their income for dirty fuel (in my state of Vermont, the poorest people living in the crummiest homes with the leakiest walls spend much of the winter heating the outdoors, because insulation and air-source heat pumps require too much up-front capital, even though over time they'd save huge amounts of money and carbon). People are immigrating to the US in large part because climate change (which they had nothing to do with) is making it impossible for them to raise food in the farms that for generations, or millennia, had provided for their families. People get sick because fracking poisons their air and water, and when they get sick they lack the money to get the health care they need. And of course solutions—even the most necessary ones—intersect with human need as well. It's not fair to condemn entire communities to impoverished oblivion because we can no longer burn the coal they once produced. These injustices focus on the most vulnerable (the one iron law of climate change is, the less you did to cause it, the sooner you suffer), and of course that means that communities of color and immigrant enclaves bear the brunt (the paddy wagon that carried me away from the Richmond protest had people who spoke six different native languages).

We've also come to recognize in a new way that the most profound injustices are being enacted on the youngest among us. I'll be dead before climate change is at its destructive peak, but people now in high school and college will be in the prime of their lives—which explains why they are leading the fight. It's good to see that the fight for a Green New Deal is led by the youngest member of Congress,

Ocasio-Cortez, millennial policy wonks like Rhiana Gunn-Wright, and the youth-led Sunrise Movement.

And it's good to see how many people are starting to line up behind them. Twenty years ago this would have seemed like a "radical" proposal, and of course that's what Fox News and Donald Trump will spend all their time branding it. But it isn't, really. (In fact, viewed in one way, it's enormously conservative: it imagines a world with levels of inequality and of carbon that resemble the ones people my age were born with. By this light, it's raising the temperature and concentrating the wealth that are dangerously radical.) I'm reminded often of one of Dr. King's favorite hymns, "Once to Every Man and Nation," which set to music a poem by James Russell Lowell. "New occasions teach new duties, time makes ancient good uncouth," it says—by which I take the idea that reality dictates what we need to do. If we get new data, then we need to respond, instead of clinging to old ways of analyzing and acting.

We have stark new information available to us: the dying of the Great Barrier Reef, the burning of the Amazon, the CO_2 monitor on the side of Mauna Loa crossing 400 parts per million and then 410, the biggest rainfalls in American history. A California city literally called Paradise literally turned into hell inside half an hour. And the stark information about inequality: the decreased life expectancy for Americans as the diseases of despair take their toll, for instance. Given this new information, we need to act, and in this case we need to act fast, because climate change is a timed test. Here Dr. King's wisdom fails us a little—loosely quoting from the Massachusetts abolitionist Theodore Parker, he used to close speeches by saying, "The arc of the moral universe is long, but it bends toward justice." That is, this may take a while but we're going to win.

The arc of the physical universe is short and bends toward

heat—if we don't win soon, then we don't win at all, because no one has a plan to refreeze the Arctic once it's melted. And winning looks a lot like the Green New Deal. Not because the kids say it's cool, and the kids are hip and woke. But because it's what reality points toward—political reality (the need to make people secure enough that we can change) and reality-reality (physics and chemistry). Given where we find ourselves, and what we've learned along the journey to this moment, it's what we have to do.

SIX

POLICIES AND PRINCIPLES OF A GREEN NEW DEAL

RHIANA GUNN-WRIGHT

People often ask me why I decided to help develop the Green New Deal. Why did I, a twentysomething black woman, think I could help develop a policy proposal to address something as big as climate change? Often, I think they expect some grand story: about incredible courage or deep ambition or a master plan for the revolution. The truth is that I was scared—and I really needed a job.

I grew up, raised by my mother and grandmother, in the same house that my mother grew up in, in a neighborhood on the South Side of Chicago called Englewood. In the thirty years between my grandparents moving in with their three babies and me being born, Englewood had gone from being a (mostly) middle-income community, close-knit and quiet, to one of the poorest, most barren parts of the city. My neighborhood had so many problems: poverty, unemployment, underfunded schools, police brutality, pollution, violence. And those were just the big ones.

I rarely saw anyone in power try to solve the problems in Englewood. And when they did try, it seemed to make things worse.

When I asked my mom and grandma why Englewood looked

like this, they didn't tell me about guns or drugs or gangs. They told me about the government. About how the highway system had been built through black neighborhoods, destroying communities that would never be rebuilt. About the public housing authority razing public housing and scattering families in the name of "urban development," only for city officials to turn around and sell the land to developers on the cheap, now that the projects sat on prime real estate. About the city underfunding black schools and then shutting them down because of "underperformance." And that's just what happened to my neighborhood—not even what happened to my family.

At the time I'm writing this, I now know that:

- My grandmother's family was not eligible for Social Security for at least fifteen years because her mother was a washerwoman, and the New Deal excluded agricultural and domestic workers (nearly all black at the time) from Social Security—President Roosevelt needed to secure votes from Southern Democrats and Southern Democrats needed cheap labor from economically vulnerable black people.
- My grandfather bought our house without any help from the GI Bill, despite being a veteran of the Korean War. My mother told me that he was too proud to apply. The truth is, pride or not, the government denied home loans to black veterans, and the notorious redlining in Chicago meant that he wouldn't have been approved anyway.
- I grew up in a frontline community—meaning that I lived in an area close to a pollution source and with high levels of air pollution. I developed asthma, like most of my friends in my neighborhood. I could barely run until I was in my late teens, and I regularly missed school, which, in turn, meant that my

> self-employed mother had to miss work. My mother and I had no idea that I was sick because of where we lived. My lungs are weakened to this day.

Progress came with a price, and the price was us. And by the time the Green New Deal came into my life, I would be damned before I paid another dime.

WHAT IS POLICY?

I have spent my life trying to rewrite systems of power, and policy is nothing if not a system for creating and distributing power. This is, of course, not how most people think of public policy. In fact, most "official" definitions of policy say something like this:

> Policy [is] a statement by government—at whatever level, in whatever form—of what it intends to do about a public problem. Such statements can be found in the Constitution, statutes, regulation, case law (that is, court decisions), agency or leadership *decisions*, or even in changes of the behavior of government officials at all levels. For example, a law that says that those caught driving while intoxicated will go to jail for up to one year is a statement of governmental policy to punish drunk drivers. The National Environmental Policy Act (NEPA) is a statement of government policy toward the environment. . . .

And: "Policy is what the government chooses to do or *not* to do" about a public problem.

This is all true. But definitions like this make policy design sound like it's orderly and contained—much like going to the doctor. You have a problem; the doctor diagnoses it; you two find the best treatment. Creating policy is more like going to the doctor with a problem, having fifteen people argue about if it's a "real" problem that requires a doctor to begin with, then having five of those people (plus some new strangers!) start arguing anew about what the cause of the problem is, only to be interrupted by the doctor's boss coming in to tell them that they can only choose two of five possible treatment options because the other three would hurt the hospital's bottom line. And once treatment begins, people argue over how to determine whether it's successful and if it should be reversed to save money or time.

Policymaking is not a science. It is a fight over whose problems get addressed, how those problems are addressed, and how public power and resources are distributed. If politics is a fight to elect people who reflect and share our values, policy is a fight to actually enact those values—to mold the world, through the work of government, into what we think it should be.

That is why, contrary to popular belief, the most important part of a policy proposal is not the details—at least at the beginning. It's the vision that the policy presents. As a statement about what the government is going to do, policy inherently tells a story about what went wrong, how the government can fix it, and who has power to shape society—whether it's the state or the public or corporations. The best policies tell compelling stories, galvanizing legislators and citizens to fight for them, and provide public servants with a clear purpose when they sit down to implement the details. The stories may shift as opponents pick new battles; the details may need tweaks or overhauls as unexpected challenges emerge. A coherent policy vision provides the foundation that both the stories and the details draw upon. Three pillars—the problem, principles, and power—form that foundation, and anchor policymaking from conception to execution.

Problems are the center of any public policy. Because policy is the government's response to a problem, policy can only be created if we agree that an issue constitutes not just *a* problem but a *public* problem—that is, a problem that affects the public that cannot be solved without the government. How we define the scope and origin of the problem determines how we'll craft a solution. That's why fossil fuel companies spend millions to sow doubts about the urgency of the climate crisis and cover up their culpability. It's not just about saving face; it's about changing our understanding of the problem and preventing government action.

Principles. Policymakers need a compass to navigate the near-infinite variety of policy designs, and principles—which include both our moral values and our theories of government—provide that compass. Remember, policymaking is collective problem-solving—not an objective "science." Policymaking, like all decision-making, is guided not only by facts but by our values—about freedom and justice, about what we deserve, about what "other people" deserve and, perhaps most crucially, about what the government should and should not do. Principles are, in short, the moral and intellectual core of a policy. They define not only how we engage with a problem but what solutions we consider at all.

Problems in our society are rooted in *power*. Asking why a problem remains unresolved leads to questions of power: Who wields it and to what end? Are the powerful negligent or malevolent? By directing and entrenching flows of government resources and attention, policy always shapes the distribution of power. Effective, lasting policy changes must change the distributions of power that led to the problem initially, or else the old malefactors will undermine any success. When selecting the mechanisms a policy will use (a loan; a new legal protection; a direct public investment; a new federal agency), policymakers are deciding how to maintain or disrupt the balance of power. And this is not limited to power in the public sector. Governments write the laws, enforce the contracts, and build the infrastructure that make a society and economy possible. Policy changes reverberate beyond the public sector into every domain of our lives.

Problems, principles, and power are the pillars of any policy vision. Together, they animate the policymaking process, guiding not just the story policymakers tell but the decisions they make about what should (or should not) be included in a given proposal.

IS THE GREEN NEW DEAL A POLICY?

The Green New Deal is a proposal for a ten-year economic mobilization to rapidly transition the US to a zero-carbon economy and, in so doing so, regenerate and reorganize the US economy in ways that significantly reduce inequality and redress legacies of systemic oppression. The congressional Green New Deal ("GND") resolution has five goals:

1. Achieve net-zero greenhouse gas emissions through a fair and just transition for all communities and workers.
2. Create millions of good, high-wage jobs and ensure prosperity and economic security for all people of the United States.
3. Invest in the infrastructure and industry of the United States to sustainably meet the challenges of the twenty-first century.
4. Secure clean air and water, climate and community resilience, healthy food, access to nature, and a sustainable environment for all.
5. Promote justice and equity by stopping current, preventing future, and repairing historic oppression of frontline and vulnerable communities, including Indigenous peoples, communities

of color, migrant communities, deindustrialized communities, depopulated rural communities, the poor, low-income workers, women, the elderly, the unhoused, people with disabilities, and youth.

The GND resolution proposes to achieve these goals in two ways. The first is through a set of "projects" that, if completed, would nearly eliminate carbon emissions in the US. The second is through a set of policies that aim to protect Americans from the disruption and instability that transitioning away from fossil fuels will create *and* reduce inequity. Some people like to refer to the first set of projects as the "Green" part of the GND and the second as the "New Deal" part. While this may be a helpful rhetorical device, it is a dangerous way to conceptualize the GND. All parts of the GND advance decarbonization—even the "non-climate" policies like universal health care, education, and job training. Similarly, the "green" projects can help reduce inequity if they are designed to create millions of well-paying jobs, bolster worker power, invest in local communities, and strengthen the social safety net—all of which the Green New Deal proposes to do.

Addressing decarbonization and inequality simultaneously has prompted critics to accuse the GND of being a "progressive wish list," not a policy. Their criticism often reveals a narrow policy vision guiding their thinking. The problem is simply the carbon in the atmosphere; Mr. Policy Doctor will prescribe the correct solution based on science; imbalances of power are mostly irrelevant, too difficult to disrupt when an urgent crisis needs solving.

This is a compelling story. But it cannot guide policymakers tasked with averting catastrophic warming, as many authors in this book show.

The Green New Deal is a new policy vision—one that will

guide government and society through the biggest task in modern history: decarbonizing our global economy within the next ten to twenty years. The stories and details of GND policy will undoubtedly change in the coming years, but they will be anchored by the vision—a conception of the problem, a set of principles, and an analysis of power—that the GND provides. Vision, however, is not enough. The GND also establishes a framework for a national economic mobilization and a set of ever-evolving and specific policies that fit within this vision and framework.

THE POLICY VISION OF THE GREEN NEW DEAL: A NEW ERA OF CLIMATE POLICY

The Problems

The Green New Deal is designed, first and foremost, to address the climate crisis at the speed, scale, and scope required to prevent catastrophic levels of warming. That is why the Green New Deal resolution has a ten-year time frame; according to the 2018 IPCC report, global emissions have to be halved by 2030, but recent reports have noted that faster reductions will likely be necessary—especially from high-emitting countries like the US.

Only the federal government wields the power to lead a national mobilization that can decarbonize the economy fast enough. But, as Naomi Klein and Ian Haney López write, we can't sustain a decade-long government-led economic transformation if the reigning ideology of market fundamentalism (described as "neoliberalism" by scholars of history, economics, and politics) and the right-wing's strategic racism define the limits of political possibility.

The power to burn fossil fuels without limit or penalty requires a political and economic system that 1) allows industrial giants to override the will of the people, who largely want this crisis stopped, and 2) condemns some peoples and places to pollution, disaster, and death. This same system has also led to declining life expectancy and rising economic and racial inequality, and has left millions without access to adequate health care, housing, and education.

These are the "problems" policymakers must address. Not only decarbonization but the ideas, systems, and inequalities that underlie the climate crisis and make sustaining decarbonization year after year a near-impossible task. This is why the Green New Deal addresses health care, housing, job security, unionization, and access to clean water and healthy food. Insecurity, division, and scarcity often prime us to resist big, unknown changes. Until we counter those threats, we can't move forward—on climate or anything else.

The other reason the Green New Deal addresses these "non-climate" issues is that a massive national effort to zero out emissions will disrupt and permanently change many lives and livelihoods.

Our current economy is predicated on a reliance on fossil fuels. They are our primary energy source. Imagine if we as humans stopped eating food and instead had to eat red algae. How would our lives change? Would we still need to eat three meals a day, or would we need to eat constantly? Where would we get the algae from? How would we grow enough to feed everyone? Who would grow it? How would we ship it? How much would it cost? What would happen to restaurants, to grocery stores? WOULD WE EVEN NEED REFRIGERATORS ANYMORE?!

Transitioning away from fossil fuels is no different. Fossil fuels do not just power our homes and cars. They power *everything*, from manufacturing clothing to streaming Netflix. Transitioning away from fossil fuels will directly affect the livelihoods of workers in the coal, oil, and gas industries *and* indirectly affect nearly every other sector. Ending our use of fossil fuels will, by its very nature, cause significant economic disruption and transformation—especially now that we have so little time. The question is simply how we will manage it.

A publicly led mobilization offers the best opportunity to design and manage the transition in ways that both grow the US economy and protect everyday people. But that is not how we've managed past mobilizations.

Every economic mobilization in American history has exploited marginalized people. The Home Owners' Loan Corporation (HOLC)—created during the New Deal to provide loans to homeowners facing foreclosure—often labeled predominantly black neighborhoods as "high risk," which discouraged lending and encouraged redlining. Today, 74 percent of the neighborhoods labeled "high risk" are low- to middle-income neighborhoods, and 64 percent are predominantly minority—meaning that these areas are still racially and economically segregated to this day. Similarly, highway expansion and urban renewal programs during the 1950s, 1960s, and 1970s displaced hundreds of thousands of residents—mostly people of color—without adequate financial assistance, erasing decades of wealth for those who owned homes and businesses. Because of this, the thought of an economic mobilization understandably frightens millions of Americans. The Green New Deal must directly address these fears, or risk losing the public support it needs to sustain itself across a decade.

Furthermore, an economic mobilization that ignores justice and equity is a danger to both marginalized people and decarbonization. For example, the racist distribution of home loans, a product of the New Deal and World War II–era policy, fueled redlining, residential segregation, *and* suburban sprawl, all of which exacerbate emissions. Similarly, many of the communities that were split open to build interstate highways never economically recovered and have since become frontline communities with high levels of pollution and degraded air quality.

The problem of climate change requires a vast, rapid mobilization, but that mobilization won't succeed if it is not just and equitable.

Principles

When it comes to climate change, you will hear people support (or refute) certain policies because it's what "science dictates." Science can help us to understand the extent of the climate crisis, identify its causes, and measure its severity. It can even suggest timelines for action. But science cannot tell us what policy solutions to pursue. That is a matter of principles.

If you cut your teeth in policymaking anytime in the past forty years of American politics, you've been surrounded by neoliberal theory presenting itself as "common sense." Thus the compass you're given to navigate the weeds of policy design is guided by a conventional wisdom soaked in the assumptions of neoliberalism: markets self-regulate, efficiently producing goods and growing the economy; government intervention is inefficient and harmful, except for the essential interventions governments make to establish and protect "free" markets; privatization, deregulation, tax cuts, and union-busting are needed for the market to function properly; inequality is a natural and acceptable consequence of capitalism. Society and government owe you nothing but the chance to work for your sustenance, and you owe nothing back.

If we try to steer a sustained national economic transformation with that compass, we will fail. Before we debate the details, our policy needs a new compass—a new set of intellectual and moral principles.

The Economic Theories Behind the Green New Deal

Several histories and theories influenced the vision of the Green New Deal, including the environmental justice movement, Keynesian economics, and the histories of World War II and the New Deal. But the most significant influence—at least during the early development of the GND—was a body of economic theory that we called the "new consensus." Exemplified by the work of economists like Ha-Joon Chang, Mariana Mazzucato, Kate Raworth, Ann Pettifor, and Joseph Stiglitz, the new consensus rejects neoliberalism as the "right" governing paradigm for modern states. Instead, it contends that many of the crises that we face are the result not of government overreach but of government's abdicating its economic responsibilities: as a market creator, as an industrial planner, and as an innovator.

Neoclassical economists believe governments should only seek to correct market failures, but as Mazzucato writes in *The Entrepreneurial State*, "that view forgets that markets are blind . . . They may neglect societal or environmental concerns . . . they often head in suboptimal, path-dependent directions that are self-reinforcing." When oil companies relentlessly pursue extraction in deeper and more dangerous parts of the earth while ignoring clean energy investment, this is "not just about market failure," Mazzucato argues, but "it's about the wrong kind of market getting stuck"; we need government "actively creating and shaping (new) markets, while regulating existing ones."

Following Keynesian theory, Green New Dealers are guided by the principle that government can and *must* do things that no other institutions do—be they the market, the church, or the family. It's the reason why strengthening the public sector and empowering the government are crucial to the policy vision of the GND, because we need a direct, efficient, and just economic transformation, which the market alone cannot—and will not—provide.

Governments have an essential, constitutive role to play in steering national economies, and the GND seeks to embrace this role by rebalancing the relationship between the public (the state) and the private (the market). This does not mean that businesses and financial markets should not or cannot have a role to play. New Consensus economists see the private sector as essential to economic transformation and prosperity. In the GND's policy vision, the government is not a handmaiden of the market, relegated to only fixing market failures; it's a leader and a risk-taker that uses its unique abilities to create, regulate, and shape markets.

With government playing a leading role in the economy, the arsenal of public powers available to policymakers is expanded, allowing us to comprehensively and equitably solve climate, social, and economic crises. Three additional principles from New Consensus economists should guide Green New Dealers as we craft policy.

First, *the US government, at all levels, must have a coordinated vision and strategy for a new "green" economy.* As Mazzucato argues, one of the reasons the US has failed to significantly reduce emissions and develop a competitive clean technology sector is that the federal government has adopted a "patchwork" approach to climate policy, uncoordinated across state agencies and jurisdictions.

Second, *public spending and investment are essential, not just for infrastructure and "public goods" but for innovation.* Federal investment in R&D funded some of our most important technologies, including GPS, nanotechnology, and many key components of smartphones. Although we have most of the technology we need for decarbonization, we still need additional breakthroughs to fully transition from fossil fuels. However, as a share of gross domestic product (GDP), public spending for R&D has declined nearly

50 percent since the 1980s. GND policies must set aside neoliberal dogmas about the need to keep public spending low and instead embrace large-scale public investment strategies that rapidly decarbonize our economy and reposition the US as a leader, instead of a laggard, in the global green economy.

Third, *the Green New Deal should invest in the real economy, not financialization.* Although engagement with financial markets and products is important to the success of the GND, its policies should invest, first and foremost, in the "real economy," meaning the parts of the economy made up of the flow of goods and services. GND policies should focus on factors that translate to real economic outcomes for everyday people (for example, job creation, wage levels, rates of unionization, and so forth) before focusing on what makes sense for financial markets and monetary policy. Monetary policy still matters, but financial markets should not be a primary concern when designing and implementing policy for the Green New Deal.

Our values informed how we interpreted the theories of New Consensus thinkers. We designed the Green New Deal as an economic mobilization not simply because it was the best solution to meet the need to rapidly decarbonize but because it was the solution that also provided the greatest opportunity for compassion, dignity, and justice. From the New Deal to World War II, economic mobilizations built the American middle class; but, as we discussed before, they only served certain Americans. With the Green New Deal, we have a chance to decarbonize, rebuild our economy, and correct those failures, but only if we design policies that do not treat injustice as a necessary—or acceptable—consequence of capitalism.

Power

The GND's vision of power is one of redistribution: from *private to public*, from *employer to worker*, from the *historically advantaged* to the *historically disadvantaged.* Unchecked burning of fossil fuels requires two things: 1) coal, oil, and gas companies with the power to circumvent or stymie any democratically imposed constraints on their business model; and 2) people and places who can be hurt, even killed, with little consequence. In short, the climate crisis can continue unabated only with immense concentrations of economic and political power.

Over the last forty years, the top 1 percent—elites, as we'll call them—have captured "more unfettered political, cultural, and intellectual power than at any point since the 1920s," Naomi Klein writes.

Neoliberal policies have allowed elites to accrue nearly all of the economic gains since 1980, a hoard of wealth they use to finance anti-labor politicians, who then pass policies to weaken unions. The result has been wage stagnation and rampant economic inequality, with declines in union membership alone accounting for about one-third of the growth in income inequality since 1972. The 78 percent of Americans who live paycheck to paycheck are rightfully anxious when a politician tells them that "saving the climate" means losing their jobs. Elites have the power to steer economic policymaking and keep working people insecure; everyday folks live at the whim of their boss and "the market."

Stagnant wages have also weakened local tax bases while increasing the need for social services, strapping state and local governments for cash and making them more vulnerable to the influence of moneyed interests. The result has been an erosion of local and community control over policy. A 2019 study of one million federal and state bills found that 10,000 of these bills were exact copies of model legislation produced by and lobbied for by elite-funded special interest groups and think tanks. Thousands of other bills included provisions from such model legislation. The majority of these bills advanced the interests of corporations and industry, even if they overrode the "will of the local voters and their elected leaders."

The success of the Green New Deal depends on the ability to reroute power away from the 1 percent and the political and economic institutions designed to serve them. If we are going to become an economy that serves people and the planet, then the people—*all* of the people—need power, and we need it now.

The Green New Deal presents a clear vision of what we need to do to address the climate crisis. But vision alone is not enough. The US has not undertaken a substantial economic mobilization in nearly eighty years, and, in the wake of neoliberalism, policymakers do not know how to design one. They need a framework, which the Green New Deal provides.

The Green New Deal as a Framework for Economic Mobilization

The only time that the United States has scaled up production at anywhere near the speed that the climate crisis requires is during economic mobilizations. The Green New Deal is, as we have seen, a proposal for a ten-year economic mobilization. But what is an economic mobilization, and why is it the best way to address the twin crises of climate change and economic insecurity?

What Is an Economic Mobilization?

An economic mobilization coordinates and deploys ("mobilizes") a nation's resources (its "economy") in response to a national crisis.

Economic mobilizations organize an economy to achieve goals that are only possible when *all* of a country's resources—public and private—are mobilized in accordance with a central common strategy and in relentless pursuit of shared objectives that supersede all other priorities. The question of whether an economic mobilization is an appropriate tactic is thus less a question of the type of crisis and more a question of the scale and nature of the crisis. To justify an economic mobilization, a crisis must be serious enough—existential, really—to demand an all-out "total war effort" from both the public and private sectors.

Given the time frame, the climate crisis—vast, existential, worsening by the day—is solvable only through an economy-wide energy transition, which requires an economic mobilization. Only a national coordinated all-out push can ramp up production of clean energy infrastructure fast enough—and ramp down emissions fast enough.

Consider the energy sector. In 2018, only 11 percent of the energy consumed in the US came from renewable sources. Electric vehicles account for less than 2 percent of all the cars sold. Buildings produce 40 percent of our nation's carbon dioxide emissions. To transition to 100 percent renewable energy, one study estimates that the US will need roughly 78 million solar panels; 485,000 wind turbines; and 48,000 solar power plants—all new—to generate approximately 95 percent of all energy from wind and solar (about 1.5 million megawatts). A utility-scale wind turbine has about 8,000 parts, many of which need to be manufactured locally, near the site of deployment. And that is only the energy sector. Similar transformations need to happen in nearly every sector of the US economy, including housing, transportation, agriculture, and manufacturing. And, as in World War II, the US has to not only equip itself but develop and produce low-carbon goods for other nations, too.

As with any industrial mobilization, it is not enough to simply produce the necessary technologies; we must also build—and manage—the infrastructure to support them. That means millions of miles of new transmission cables to support "smart" electric grids that can integrate renewables; thousands of new charging stations for electric vehicles; and new manufacturing facilities to produce electric furnaces, heaters, and stoves to equip our homes and businesses—and that is still the beginning. We also need new practices to replace carbon-intensive industrial and agricultural processes. Regenerative agriculture, ocean farms, and even electrolysis—we need all of it to zero out emissions, and we need it fast.

An industrial mobilization can make the rapid, vast transformation we need possible. This is due, in part, to the policy coordination they generate. There are few other times when all of the levers of government—including regulation, legislation, executive action, and procurement—are aligned, coordinated, and leveraged toward a single set of goals. But it is also due, in large part, to the unprecedented levels of public investment they unleash, especially in service to initiatives that are too "risky" or big to attract private investors.

Critics who doubt our nation's capacity to achieve a transition of the scale and speed the Green New Deal proposes should heed the lessons of the World War II mobilization: set the production targets you need to win, even if they seem impossible at the outset, and then hustle to meet those targets through massive, coordinated, and strategic public investment and collaborations with private industry.

Lessons from the World War II Mobilization

The World War II years are an illustration of the deep cooperation between the public and private sector that a successful economic mobilization requires. The WWII economic mobilization was designed not only to put the US on war footing but also to produce weapons, artillery, and machinery—the "arsenal of democracy"—that Allied forces needed. In his 1942 Annual Message to Congress, FDR set ambitious production targets: 125,000 planes, 75,000 tanks, 35,000 anti-aircraft guns, and 10 million tons of merchant ships in 1943 alone. "Our task is hard—our task is unprecedented—and the time is short," Roosevelt said after announcing the new goals.

These were impossible targets when compared with US production capacity before WWII. Before the war, a year's worth of production built enough ships to carry a total of 0.3 million deadweight tons and made 20.3 million pounds of airframes for plane construction; fewer than 100 tanks and 3,700 planes were made annually in prewar America. The mobilization transformed our production capacity. At the peak of ship production in 1943, the US built enough vessels to carry 18 million deadweight tons in a single year. Airframe production peaked the next year—787.1 million pounds built in 1944 alone. By the end of the war, the US had produced 299,293 aircraft and 88,410 tanks.

The production of component materials also skyrocketed, with the production of synthetic rubber, magnesium, and aluminum increasing anywhere from 7 times to 288 times the prewar average. Unemployment plummeted as a result of the industrial boom, dropping from 14.5 percent in 1940 to 1.6 percent in 1945—all while increasing wages and significantly reducing income inequality.

Tightly coordinated public-private partnerships and high levels of public investment made it possible for the US to rapidly increase production capacity in just a few years. The decision to prioritize public-private partnerships and support them with large infusions of public capital was a matter of sheer material need. The US had to equip at least 12 million American troops and also supply the British and French militaries. There was simply no way to manufacture that much materiel solely through public means. The government had to find a way to mobilize the US consumer economy—the largest industrialized economy in the world—to make military goods. The United States spent nearly $20 billion—almost 10 percent of all WWII defense spending—on manufacturing and machinery, and much of that investment went toward building "government-owned, contractor-operated" ("GOCO") facilities: publicly owned factories run by privately owned businesses. The US government invested so much money in building new manufacturing facilities that by the end of the war it owned "close to a quarter of the nominal value of all of the nation's factories."

The success of public investment and the large network of GOCO factories would not have been possible without the involvement of business. Not only did business owners convert civilian factories for wartime production—going so far as to retool auto and radiator factories to produce tanks and helmets—they also lent their expertise and experience. Private engineers partnered with military experts to develop and upgrade new weapons. Corporate executives joined the War Department and traveled the country, recruiting other business leaders to commit to military production. Perhaps most importantly, civilian contractors worked with military officials to manage complex supply chains, train new workers quickly, and run increasingly large and complex production facilities.

Critics argue that economic mobilizations—especially wartime mobilizations—inherently stifle "free enterprise" and disadvantage the private sector. But private industry benefited from the World War II mobilization. Companies that operated GOCO facilities or manufactured war goods received contracts that reimbursed them for all authorized production costs and guaranteed profit. The US government also paid private firms—especially smaller firms—to license new and patented technology so that it could be produced at scale.

National economic mobilization requires huge workforces,

which in turn tighten the labor markets, benefit workers, and create opportunities to balance economic inequities much more quickly than in normal economic circumstances. During WWII, for example, the share of the national income that went to the top 1 percent dropped from 15 percent in 1939 to 10 percent in 1945, and it fell even more steeply—13 percentage points—for the top 10 percent. The mobilization especially benefited marginalized workers, as the share of national income for many—including African Americans, women, and agricultural workers—rose sharply during the 1940s.

Economic Mobilization as a Response to the Climate Crisis

Economic mobilizations free the public sector to invest significant capital directly into private industries in ways that are coordinated, targeted, and strategic. During mobilizations, the government "absorbs" risk for new long-term (or large-scale) projects—by guaranteeing loans, providing early-stage capital, and so on—which makes it possible to develop new products without having to wait on private capital that is short-term and risk-averse. That risk-taking is crucial for the climate crisis. Although the US already has most of the technology we need to decarbonize, we still need some new breakthroughs—particularly when it comes to sectors like aviation and shipping. Economic mobilization gives the public sector a green light to invest heavily and directly in necessary R&D *and* to coordinate private capital to support these projects, once they have been proven to work.

In addition to unleashing public investment, mobilizations also amplify the public sector's ability to strengthen and stabilize demand. Essential low-carbon industries, like solar and wind, have struggled to gain a foothold in the US, due in large part to inconsistent and insufficient demand. An economic mobilization allows the government—as both customer and regulator—to expand demand for "green" goods and reshape markets to support them.

Take solar energy as an example. To fully transition the US energy system to renewable sources, about 57 percent of all residential rooftops that are suitable for solar installations need to have solar installations. But in 2020, an average-sized residential solar installation will cost between $11,400 and $15,000 *after* solar tax credits—far more than most American homeowners can afford. Much of the expense is due not to the price of the hardware but to "soft costs" related to marketing, customer acquisition, and navigating different layers of permitting. It is a vicious cycle: fewer homeowners install solar because of the high cost; the price of residential solar does not decrease as fast because "soft costs" remain high; "soft costs" can't go down until it becomes easier to acquire customers and deal with permitting. Until now, federal and state governments have done little to help, outside of offering tax credits, which do very little to defray upfront costs and thus do little to increase demand for solar installation from homeowners.

Green New Deal policies, however, can permanently increase demand by requiring new construction to be zero-carbon, funding solar retrofits of existing buildings, and streamlining permitting across states. Solar manufacturers no longer need to spend money marketing to hesitant customers—every new home needs solar panels. Instead of buying more ads, manufacturers can invest more in scaling up production to meet the increased demand, and the scaling up of production will spur new innovations that drive down hardware costs even further. Meanwhile, property owners can get grants from newly established green banks, expanding demand for solar to homeowners who otherwise cannot afford the upfront cost. This would employ thousands of people: there are well-paid public sector jobs waiting to be created on your house's bare roof.

Economic mobilizations also present the possibility of reexamining and renegotiating our social contract—to decide what kind of country, what kind of society, we want to carry into the future. Economies and societies do not exist separately. In mobilizations—especially ones as massive as the one proposed by the Green New Deal—we need social policy that serves the goals and principles of the GND *and* provides the support necessary to maintain an economy at nearly full employment. That requires changes not only to labor policy but to every part of our social safety net, from healthcare policy and childcare policy to workforce policy and housing policy. Some of the most significant economic effects of the original New Deal came from new social policies—like Social Security—and public employment programs that employed hundreds of thousands of Americans, which in turn maintained and prepared the workforce necessary for the production boom of WWII.

Mobilizations also create tight labor markets that redistribute wealth and reduce inequality. In fact, mobilizations distribute wealth far more effectively than redistributing wealth through the tax system, particularly for low-income and marginalized workers. But they are not inherently just—especially if not everyone can access the jobs they create. How can we attract enough workers for a national climate mobilization if the average cost for a year of childcare ranges from $5,500 to $25,000? How can families move to better-paying jobs in the mobilization if they are dependent on their employer-sponsored health care? How can people reenter the labor force if they do not know where to go for training or job placement? They can't—unless we meet those needs through the public sector. That is why the Green New Deal includes commitments to a federal jobs guarantee, universal childcare and health care, and significant investments in education and workforce education.

These commitments were designed to ensure that any mobilization to address climate change adheres to the principles of the Green New Deal from design to implementation. The price of national progress cannot be systematized oppression—not again.

The Green New Deal as Public Policy

Alexandria Ocasio-Cortez has described the Green New Deal resolution as a "request for proposals," and sure enough, communities and leaders from across the country have begun to translate the vision and framework of the resolution into concrete policy proposals. Although still nascent and evolving, these GND policies share four characteristics that differentiate them from traditional climate and economic policies.

First, all GND policy, whether narrow or broad, serves a triple bottom line: achieve the decarbonization goals set out by H.R. 109, reduce income inequality, and redress systemic oppression. The Green New Deal for Public Housing, for example, will retrofit and upgrade 1.2 million federally managed homes, reducing their carbon emissions while also creating hundreds of thousands of jobs. Grants for paid workforce development programs will train public housing residents and prepare them for the 250,000 jobs—all paid at prevailing wage levels—that the bill creates. All grant applications from local organizations must also be approved by resident councils, giving residents unprecedented control over how money invested in their homes will be spent. New York's Climate Leadership and Community Protection Act aims for a triple bottom line with equitable investment, mandating New York reach net-zero greenhouse gas emissions; establishes intermediate decarbonization goals—including 70 percent renewable energy by 2030; and requires that marginalized communities receive no less than 35 percent of the bill's investment.

Second, GND policy works to shape markets and create demand so that low-carbon and no-carbon goods become the default, rather than the alternative to carbon-intensive goods. Maine's Green New Deal and Los Angeles's Green New Deal include renewable portfolio standards that require 80 percent of all energy be generated from renewable sources by 2040 and 2036, respectively, which will significantly increase the demand for renewable energy. New York City's Climate Mobilization Act achieves a similar shift in demand and energy markets but through a different mechanism. By altering the city's building code to include emissions caps for medium-sized and large buildings, requiring all new residential and commercial buildings to include either green roofs or solar installations, and adding wind to the Department of Buildings' "toolbox" of allowable renewable energy technologies, the bill greatly expands markets for low-carbon building materials, renewable energy, and related technology, in New York and nationally.

Third, GND policy mobilizes public investment for sector-wide decarbonization, while ensuring that the investment provides workers, marginalized populations, and vulnerable communities with both a path into the new economy and protection from disaster. For example, the Green New Deal plan put forward by Senator Elizabeth Warren would invest $10 trillion—public and private—over ten years, with $2 trillion going toward green manufacturing and research; $15.5 billion toward sustainable agriculture and localizing food systems; and at least $1 trillion to frontline and fence-line communities. The Green New Deal plan from Senator Bernie Sanders would invest $16 trillion, including $2 trillion toward renewable energy and modernizing our electric grid; approximately $3 trillion to weatherize and retrofit low- and middle-income homes and small businesses; and roughly $2.7 trillion to help working-class families purchase electric vehicles. Sanders's plan would also invest

$40 billion in a Climate Justice Resiliency Fund that would be used for projects as varied as community centers and shelters with reliable backup power on one hand, and, on the other, wetland restoration and climate-related adaptation for frontline communities.

Finally, GND policy works to build power within and among those who are marginalized by prioritizing these communities in funding, policy design, and implementation, enabling local control whenever possible. The GND resolution requires that democratic processes, "inclusive of and led by frontline and vulnerable communities and workers," be used to "plan, implement, and administer" the Green New Deal at the local level. So far, legislators have listened.

With the exception of the Climate Mobilization Act, all of the legislation discussed in this section positions frontline communities as drivers of climate policy—whether these are environmental justice working groups, transition advisory councils, or commissions on a just transition. New York's environmental justice working group "will establish criteria to identify disadvantaged communities for the purposes of co-pollutant reductions, greenhouse gas emissions reductions . . . and the allocation of investments" that result from the Climate Leadership and Community Protection Act. Similarly, at the federal level, the Climate Equity Act introduced by Senator Kamala Harris and Representative Alexandria Ocasio-Cortez would establish a Climate and Environmental Equity Office, install a senior adviser for Climate and Environmental Justice in sixteen federal agencies, and create "an equity screen" for all federal rules or regulations related to climate.

We've barely begun the long journey to a fair, fossil-free economy, but already the policies of the Green New Deal offer myriad designs and mechanisms to achieve its goals: municipal codes, emissions caps and timelines, local control, federal grants to community organizations, and sector-by-sector investment and transformation. The details will keep changing as we learn how best to decarbonize equitably and mobilize the American people—our hands, our creativity, our resources—to remake our economy, while caring for one another every step of the way. But no matter what we encounter in the weeds of policy blueprints and implementation, the vision of the Green New Deal provides the compass we'll need.

SEVEN

THE ECONOMIC CASE FOR A GREEN NEW DEAL

JOSEPH STIGLITZ

THE VISION OF THE GREEN NEW DEAL

When a country is in crisis, bold action is required. In 1933, the US faced a crisis. The country had been mired in the Great Depression for four years, and it seemed to be getting worse every day. A quarter of the population was out of work. Farmers' incomes had fallen by half or more. President Franklin Delano Roosevelt understood the urgency of the moment, because he understood the mistakes of the past: The country had delayed doing anything during the presidency of Herbert Hoover. If Hoover had acted promptly and appropriately after the stock market collapse of 1929, who knows how much suffering might have been avoided? But he didn't, and FDR came up with a grand vision of how to restore prosperity: the New Deal.

So too, when the country faced the threat of fascism, we mobilized all our resources to fight the enemy and preserve our values.

Both of these situations provided opportunities that went well beyond meeting the exigencies of the moment. In the 1930s, we passed social legislation (including Social Security) and labor legislation (including the Wagner Act) that transformed our society.

During World War II we went from an agricultural economy and a largely rural society to a manufacturing economy and a largely urban society. The temporary liberation of women as they entered the labor force to meet the country's war needs had long-term effects.

This is the ambition, a realistic one, of the Green New Deal. We are facing a climate crisis, a crisis that requires a bold, forceful, and comprehensive response of the kind envisioned by the New Deal; and the Green New Deal could bring about a twenty-first-century societal transformation that would create a more inclusive society. There is absolutely no reason the innovative and green economy of the twenty-first century must follow the economic and social models of the twentieth-century manufacturing economy based on fossil fuels, just as there was no reason that that economy had to follow the economic and social models of the agrarian and rural economies of earlier centuries.

The Green New Deal is urgently needed, affordable, and would benefit our economy.

Why a Green New Deal Is Necessary

Advocates of the Green New Deal say we must deal with climate change *now*, and highlight the scale and scope of what is required to combat it. They are right.

We have known about the threat of climate change for decades, but our steps to minimize or stop it have been halting and clearly insufficient. Carbon emissions and concentration in the atmosphere have increased enormously. Taking carbon out of the atmosphere is far more expensive than not releasing it in the first place, and largely unproven at scale. The most cost-effective way to stabilize the climate at 2°C is to cut emissions now rather than relying on negative emissions later.

That's why it is imperative that we change our economic system to a green economy. It's not just about the amount of electricity we use, or how we produce that electricity, it's about every aspect of our economy—housing, transportation systems, the structures of cities, and what we consume, including what we eat. The modern industrial economy of the nineteenth and twentieth centuries was built on energy derived from fossil fuels. The twenty-first century innovation and service-sector economy will have to be built on energy derived from renewable resources like wind and solar. And it will have to do a far better job economizing on energy.

The market won't be able to make this transformation on its own. Government will need to play an important role in a society-wide mobilization: making necessary, immediate public investments (including building out green infrastructure and mass public transit), and implementing and enforcing regulations and appropriate environmental pricing.

The magnitude of the task is enormous, and an even better analogy than the New Deal might be the mobilization to fight World War II, when the entire nation worked together to save the country. But while the task is enormous, it is both doable and affordable.

THE COSTS OF DOING NOTHING: WHY WE CAN'T AFFORD NOT TO MOBILIZE FOR A GREEN NEW DEAL

The United States is already experiencing the financial costs of ignoring climate change—in recent years the country has lost almost 2 percent of GDP in weather-related disasters including floods, hurricanes, and forest fires. We're going to be paying for climate change one way or another; this is one of those situations in which the aphorism "an ounce of prevention is worth a pound of cure" truly applies.

The costs of doing nothing are greater than the costs associated with controlling greenhouse gas emissions to rein in climate change.

While no single weather event can necessarily be attributed to climate change, the increased *incidence* of such events clearly can. Recall that one aspect of climate change is the growing variability of the weather, so the extreme cold spells during the winter of 2019 can also be chalked up to climate change. Crops can be decimated by just one episode of below-zero temperatures, so the costs of this weather seesaw can be enormous.

The toll on human health from climate-related diseases is just being tabulated, but treating these conditions, too, will run into the tens of billions of dollars—not to mention the as yet uncounted number of lives lost.

Around the world, there will be climate refugees—people no longer able to earn a living on their land, either because it has become submerged or because it has become inhospitable to agriculture and people—and species will be destroyed. New conflicts will arise, with political repercussions impossible to predict. In short, the standard cost-benefit analysis can only provide a glimpse of the possible global societal cost of climate change.

UNDERESTIMATING THE COSTS OF THE CRISIS

In considering our possible climate futures, an old maxim comes to mind: "Prepare for the worst." In technical terms, this idea is sometimes called the precautionary principle. In terms of standard economic language, we are risk-averse. That's why people buy insurance.

Unfortunately, too much of the economic research on climate change has been produced by rosy optimists who suggest that we should be content with accepting an increase in temperature of 3.5 degrees Celsius—a number far higher than the 1.5 to 2°C agreed on in the Paris and Copenhagen accords. These optimists want us to gamble with our only planet by allowing a carbon-heavy atmosphere not seen for more than 3 million years, well before the beginning of the human era.

There are so many flaws in the models employed by these analysts that it is hard to take the models seriously; but given that they have proven useful to people who would do nothing about our climate crisis, let me dwell for a moment on several of the critical errors.

First, their models fail to recognize the complex feedback mechanisms, nonlinear effects, and tipping points in our climate system. Naomi Oreskes and Nicholas Stern write, as an example of the complex, cascading disruptions that economic models rarely capture, that "a sudden rapid loss of Greenland or West Antarctic land ice could lead to much higher sea levels and storm surges, which would contaminate water supplies, destroy coastal cities, force out their residents, and cause turmoil and conflict."

As a result, these models of climate change pay insufficient attention to the real possibility of very adverse outcomes. The greater the weight we assign to such very bad outcomes, the more precautions we should take. By assigning little weight—far too little weight—to very adverse outcomes, these studies systematically *bias* the analysis against doing anything.

It is precisely when the consequences of climate change are large that we are least able to absorb the costs. There's no insurance fund that we can buy into and draw on should we need massive investments to respond to large increases in sea level rise and other extremes in the "fat tail" of unmitigated climate change. The fact is that in these circumstances our world will be poorer, and less able to absorb such losses.

Instead, those economists who blithely say we should accept a world warmed by 3.5 degrees Celsius—a world with widespread famine and whole regions of the earth made uninhabitable by direct heat—project continued growth little altered by these cataclysmic changes, if altered at all. Their models are systematically underestimating the future risks of climate breakdown while overestimating the stability of our economy in a warming world.

In the end, though, responding to climate change is about more than just a cold cost-benefit analysis, or how to value the lives lost or the species destroyed should we continue on the current course. Indeed, some people argue that looking at the consequences of the destabilization of the climate in terms of GDP can be a distraction, because so much more is at stake. A proper response involves assessing the value of life, the costs of strife, and the disruption of communities. We are on track for a future in which there is a significant risk that a peaceful and prosperous human civilization is difficult, if not impossible.

CAN WE AFFORD IT?

Critics ask, "Can we afford a Green New Deal?" They complain that the policies are unreasonably expensive, and that Green New Deal proponents confound the fight to preserve the planet, to which all right-minded individuals should agree, with a more controversial agenda for societal transformation. They are wrong on both counts.

Even the way the question is posed is wrong: When the US was attacked in 1941, no one asked, "Can we afford to fight the war?" It was an existential matter. We could not afford *not* to fight it. The same is true for climate change. Moreover, as we have already noted, we will pay for climate change one way or another, so it makes sense to spend money now to reduce emissions rather than pay a lot more later to manage the consequences, not just from the weather but also from rising sea levels.

In short, we *must* afford it. So much is at stake now, just as it was in World War II. Climate change is our World War III.

The question we really should be asking is, "Why are we postponing the economic benefits of a Green New Deal?" By delaying the investment needed to tackle the climate crisis, we are burdening ourselves with real costs in rising damages.

A Surfeit of Resources: The Savings Glut and Unemployment

Delay seems especially foolish, given the extent to which resources are underutilized globally.

Half the world seems worried about the lack of resources to fight climate change and the other half about a superabundance of resources—too much labor, evidenced by growing unemployment (for example, as a result of robotization and AI) and a savings glut (too much savings relative to investment opportunities). It cannot be true both that we have a surfeit of resources and that we can't afford a green transition.

In fact, making the investments necessary for a quick transition to a green economy is a good way to ensure that we don't confront high levels of unemployment anytime in the immediate future. Many of the actions required, such as insulating homes and installing solar panels, call for skills that are easily learned. The current modest efforts to fight climate change can be quickly scaled up.

The seeming contradiction between a savings glut and huge unsatisfied investment needs raises numerous theoretical and practical questions.

First, how, as economists, can we explain this quandary? Aren't markets supposed to work so that demand equals supply, so that there would never be a savings glut (or, for that matter, unemployment)?

The underlying problem can be traced to a fundamental weakness in financial markets: much of the savings is long-term—people saving for their retirement, sovereign wealth funds holding money for future generations, or university and foundation endowments. Much of the needed investment is also long-term. But standing between the two are our short-term-focused financial markets, for which thinking two quarters ahead is considered long-term thinking.

Second, are there policy "solutions," or at least changes that would ameliorate the problem?

Changing the tax laws and corporate governance would help encourage long-term thinking: for example, increasing the holding period before CEOs can cash in their stock options, or giving more weight to long-term investors. So would requiring trustees of pension funds and foundations to take a long-term perspective. If firms, pension funds, foundations, and other organizations took a long-term perspective, they would have considered the consequences for and of climate change. Making corporations account for their climate risk would focus more attention on the consequences of *not* doing enough about climate change.

Third, asserting that society as a whole has enough resources for a Green New Deal is one thing; explaining how households, firms, and government can actually finance these investments is another. How is that to be done?

One way is through the creation of a national green bank. This kind of bank would provide funding for investment, including to the private sector, for climate change—to homeowners who want to make the high-return investments in insulation, or to businesses that want to retrofit their plants and headquarters for the green economy.

Such "development banks" can also help design large-scale projects and construct financial instruments that diversify risk. Development banks have played an important role in developing countries and emerging markets. They can also play an important role in the green transition of the United States and the world.

Shifting Resources and Reforming the Economy

While there are ample underutilized resources to harness in the war to preserve our planet, almost surely there will have to be a redeployment of resources to win this war. During World War II, bringing women into the labor force expanded productive capacity. We moved underemployed and inefficiently employed people from the rural sector. And we moved production from making consumer goods (like cars) to making the items required for military operations.

Similarly, some of the changes in the workplace that would result in a more equitable economy and society would enhance the efficiency and productivity of the economy, better enabling us to address climate change. For instance, greater flexibility in hours would enable more women to participate in the labor force.

Some steps to fight climate change will be easy. For instance, we can eliminate the tens of billions of dollars of fossil fuel subsidies and move resources from producing dirty energy to producing clean energy. Redesigning cities to be more energy-efficient (for instance, with better public transportation) will also result in their being more time-efficient.

Can the Government Afford a Green New Deal?

Eliminating fossil fuel subsidies, increasing the government's energy efficiency, and turning to a green bank for funding much of the required investment may not suffice: The government may still need additional funds. You could say, though, that America is lucky. We have a poorly designed, regressive tax system rife with loopholes—so it would be easy to raise money while we increase economic efficiency. Taxing dirty industries, imposing a broad range of taxes on pollution and on destabilizing short-term financial transactions, closing tax loopholes, including those that allow highly profitable corporations to get away with paying almost no taxes, ensuring that capital pays at least as high a tax rate as the people who work for a living—all of it would provide trillions of dollars to the government over the next ten years, money that could be spent to fight climate change. And many of these taxes (those on pollution and financial transactions, for example) would actually improve the performance of the economy.

Standard estimates, though, are almost surely conservative, since they do not consider how the GND would stimulate the economy on both the demand and the supply side, thereby generating still more tax revenue.

I suspect that between the stimulus to growth that the GND would provide (described more fully below), and the redeployment of resources and tax measures discussed above, there would be more than sufficient resources to win the climate war.

Some, however, suggest that government may need to borrow. Again, the good news is that deficit fetishism seems finally to be defeated, as even mainstream economists such as Olivier Blanchard have argued that there is scope for government to undertake more debt.

One should never look at the liability side of a balance sheet, whether that balance sheet belongs to an enterprise or the government. If debt increases in order to finance a highly productive investment, the government or the enterprise is better off. And it should be obvious that if the economic growth is greater than the real interest rate—clearly the situation now—the burden of servicing the debt will diminish over time. Both of these help explain why the US debt-to-GDP ratio of some 118 percent at the end of World War II proved to be no problem: we continued to invest, the economy continued to grow, and before long, the debt-to-GDP ratio had shrunk to around 55 percent by 1955 and below 40 percent by 1964.

THE ECONOMIC BENEFITS OF THE WAR ON CLIMATE CHANGE

In fact, the Green New Deal is not only affordable and not only would we save money by acting on it now but it would actually be *good* for the economy. The war on climate change, if correctly waged, would benefit the country in the same way that World War II set the stage for America's golden economic era, with the fastest rate of growth in its history amid shared prosperity.

The Green New Deal would stimulate demand, ensuring that all available resources were used. It is only in periods of high aggregate demand (such as the late 1990s) that marginalized groups have been effectively brought into the labor force. Reducing carbon emissions, if done correctly, would be a great job creator as the economy prepares itself for a world with renewable energy. The investments required to retrofit the economy by themselves would generate large numbers of jobs. Tightening labor markets will also raise wages, which have long been nearly stagnant. As this book goes to press, COVID-19 is wreaking economic havoc and tens of millions of

Americans are losing their jobs. Congress is already spending trillions of dollars for stimulus and relief efforts. Recovery policies can help rescue the economy now—bringing down unemployment and restarting our economic engine—at the same time that they help create the kind of economy we need for the future.

With well-designed family leave and support policies and more time flexibility in the labor market, we could bring more women, more people of color, and more of the elderly who are able and want to work into the labor force.

Because of our long legacy of discrimination, many talented people are not able to fully use—or use at all—their skills in our economy. Together with better education and health policies and more investment in infrastructure and technology—true supply-side policies—the productive capacity of the economy would increase, providing some of the vital resources the economy needs to fight and adapt to climate change. But beyond this, these policies would be good for these individuals and their families and would begin to address one of the country's real blights—its deep and pervasive inequalities.

The transition to the green economy would, moreover, likely usher in a new innovation boom. Moving to a green economy has already spurred a lot of innovation; taking it further would stimulate even more. The energy innovations of the nineteenth and twentieth centuries were key to increasing standards of living in the era of fossil fuel. These new innovations will be key to preserving and enhancing standards of living in the twenty-first century. Indeed, parts of the world are suffocating on fossil fuel emissions; unless alternatives are found, life expectancies and health will deteriorate significantly—even without looking at the impact of climate change.

Together, the combination of demand- and supply-side effects of the Green New Deal could be a transformative moment for our economy and our society—in the same way that the New Deal and World War II were more than three-quarters of a century ago.

Strengthening Our Democracy

Lastly, the grassroots movement behind the Green New Deal offers a counterpoint to the pervasive cynicism and despair that have contributed to the wave of populism and nativism sweeping the world. Adopting the Green New Deal would renew faith in our democracy: A grassroots movement can move us toward an alternative to an economic model that places society on a collision course with nature. And it can do so while promoting inclusion and societal well-being.

EIGHT

A GREEN NEW DEAL FOR THE GULF SOUTH

COLETTE PICHON BATTLE

WHY A GREEN NEW DEAL FOR THE GULF SOUTH?

The Green New Deal is a visionary plan for addressing the climate crisis and investing in national infrastructure through the creation of millions of living-wage jobs. It offers an opportunity to heal our communities and protect generations to come.

Here in the Gulf South, we know climate disaster. Hurricane Katrina changed my life. I moved back home, to Slidell, Louisiana, in 2006. I realized my community needed lawyers—someone to read all the papers a disaster creates. They were being asked, in the middle of trauma, to sign away their rights. I'm only the third lawyer to come from my community. So I read the papers, and I decided to stay. It was about two years after Katrina that I first saw the Louisiana flood maps, which show historical and impending land loss due to sea level rise. On that particular day, at a community meeting, these maps were used to explain how the 30-foot tidal surge that accompanied Hurricane Katrina had flooded communities like mine in south Louisiana and across the Mississippi and Alabama coast. It turns out that over the last fifty years, we had lost land that had previously been our buffer from the sea.

I volunteered to interact with the graphics on the wall, and in an instant my life changed for the second time in two years. The graphics showed the disappearance of my community and many other communities before the end of the century. Land, trees, marsh, bayous, friends, neighbors, family. I had assumed that it would always be there, as it had been for thousands of years. I was wrong.

There in that community center, standing with other black, native, and poor residents of south Louisiana, I ceased to be bound to them solely by the needs of short-term disaster recovery—I was now also bound to them by the task of ensuring that our communities would not be erased by sea level rise due to climate change.

Our firsthand experiences in the Gulf South—Texas, Louisiana, Mississippi, Alabama, and Florida—affirm the call of scientists to rapidly address climate change and make frontline communities more resilient in the process. The Green New Deal resolution calls for "the use of democratic and participatory processes that are inclusive of and led by frontline and vulnerable communities and workers to plan, implement, and administer the Green New Deal mobilization at the local level." Gulf South communities are taking that call seriously by proactively defining what the Green New Deal can and must do for our region, in preparation for leading the implementation effort when federal and state policy is passed.

We are coordinating across the region through "Gulf South for a Green New Deal," an organizing initiative and policy platform with support from over one hundred organizations across the five Gulf South states. Our work proves that media depictions of the Green New Deal as a program for liberal elites could not be further from the truth. Here on the third coast, poor black, white, brown, and native people, small businesses, neighborhood associations, and regular folks from all walks of life are ready for a Green New Deal, and we know the same is true of people all across this country.

THE GULF SOUTH: CONTEXT AND PERSPECTIVE

Texas, Louisiana, Mississippi, Alabama, and Florida comprise the US Gulf South. Connected by the Gulf of Mexico, the Gulf South has deep, historic ties to the Caribbean and Central America. Together these states span America's Southwest, the Deep South, the Bible Belt, the Black Belt, and the Atlantic coast. The region plays a pivotal role in the US economy, the energy industry, the current national defense infrastructure, and the ongoing global advancement of liberty, science, social movements, and social innovation.

Social movements originating in Gulf South states have given rise to both historic and modern struggles for human rights and civil rights around the globe. W. E. B. Du Bois characterized the US South as a key link in the chain for the American working-class fight against exploitation and violent suppression. "As goes the South, so goes the nation" was a truism of his time and remains a statement of fact today. And while the political South has had a dark history, southern political resistance in the five Gulf South states has proven some of the most effective uses of our modern democracy.

A Gulf South economy first built by enslaved labor has transitioned to become the epicenter of the oil and gas industry. In Louisiana alone, there are 125,000 miles of pipeline, and the state hosts nearly half of the country's refineries. Texas refines 41 percent of the nation's oil and gas and produces more crude oil than any other state. The Gulf Coast is home to 45 percent of US petroleum refining capacity and 51 percent of US natural gas processing plant capacity. This toxic economy, its foundations constructed by slaves, is now fueled by environmental racism.

Due to the overwhelming political power of the fossil fuel lobby, laws protecting the workers who actually extract and refine the oil and gas are exceptionally weak in the Gulf South. Alabama, Florida, and Louisiana are at-will states, where workers can be fired without cause or notice, and all five Gulf South states harbor right-to-work laws, which, according to the AFL-CIO, "tilt the balance toward big corporations and further rig the system at the expense of working families [and] make it harder for working people to form unions and collectively bargain for better wages, benefits and working conditions." Without protections, workers are vulnerable to unsafe industry practices that endanger them. The British Petroleum oil drilling disaster of 2010, when the BP-leased and Transocean-owned rig *Deepwater Horizon* exploded, releasing nearly 5 million barrels of oil into the Gulf of Mexico, is remembered for its ecological impact, but less remembered are the eleven oil rig workers who perished.

Communities, like workers, suffer for the abuses of the fossil fuel industry. The BP disaster released more than 200 million gallons of oil into fragile marine ecosystems in two countries, ecosystems that were the lifeblood of the local economy. Tens of thousands of seafood harvesters, tourism companies, and hospitality workers were out of work for years, and many never recovered. The oil drilling disaster also had direct health impacts on cleanup crews and gulf residents because of the use of toxic chemical dispersants. These chemicals dispersants are banned in Europe due to known harm to humans yet remain approved for use in Gulf South communities by government and industry.

This same oil and gas industry in the Gulf South poses a toxic threat to frontline communities during climate disasters. During Hurricane Harvey in 2017, millions of pounds of toxic pollutants were leaked into the air by industry. Though production of oil has decreased on the Gulf Coast, toxic chemical production is flourishing. Like oil and gas operations, a disproportionate amount of chemical storage and refining occurs in low-income and predominantly black and brown communities. In addition to targeting marginalized communities, the multifaceted petrochemical industry emits excessive quantities of greenhouse gases, causing an acceleration of the global climate crisis.

Of course, our coastal location and low-lying land also makes the Gulf South ground zero for this nation's climate disasters. By 2030, the number of flood events in the region is expected to double, putting nearly 2 million homes at risk in just Florida and Texas alone. Louisiana is experiencing one of the highest relative levels of land loss to sea level rise on the planet. The acceleration of land loss is due to a deadly combination of sea level rise, oil and gas operations, and engineered federal controls over the natural flow of the Mississippi River. Gulf Coast communities are significantly disadvantaged in preventing and recovering from flooding as a result of social and economic inequalities. I saw after Katrina how the most severely impacted areas are host to long-standing tribal territories and historic black communities.

In addition to extreme weather, the Gulf South faces additional threats to the health and resilience of coastal communities due to rising temperatures and increased tidal flooding and other effects of sea level rise. The region is already experiencing dangerous heat waves, which will only continue to worsen if immediate action is not taken. The number of days exceeding 100 degrees is rising in all five states and is expected to quadruple by 2050.

GULF SOUTH FOR A GREEN NEW DEAL

Gulf South for a Green New Deal was launched in New Orleans in May 2019 with more than 800 advocates—farmers, fisherfolk, and community leaders from across the Gulf South. The launch was followed by a five-state process of formalizing frontline voices. In November 2019, the Gulf South for a Green New Deal Policy Platform was formalized.

In crafting a Green New Deal for the Gulf South, we began with an assertion of fundamental human rights—every person has an equal right to clean air, water, economic security, and self-determination. Every person matters, and every voice matters. This informs how we approached the bottom-up platform design process. We asked people from all corners of society what they needed and what it would take to meet *all* of our needs, with nobody left behind.

Dr. Beverly Wright of the Deep South Center for Environmental Justice taught us about environmental justice. The floodwaters leave no doubt that to protect everybody's right to a stable climate, we must ban new pipelines and drilling leases for extractive industries on Gulf South lands and territorial waters, remediate toxic pollution to the satisfaction of impacted communities, and ramp up investment and job opportunities in nonextractive economies. New renewable-energy job opportunities must prioritize opportunity and access for oil and gas workers and fence-line communities currently reliant on extractive industry jobs.

Physical infrastructure can also be redeployed in the service of the energy transition. We can take advantage of decommissioned oil rigs and other leftover oil and gas infrastructure for reuse in the advancement of renewable energy infrastructure.

Policies aimed at reducing greenhouse gas emissions must also clean up the toxic lands left behind by oil and gas, with the same financial support and urgency. In this cleanup, and in every other aspect of the fair and just transition, the costs and burden must be paid by the industries that have polluted communities and deliberately deceived the public about the threat their operations pose to our climate.

Looking beyond the fossil fuel industry, we found that workers in many other professions are also eager to play their part in a Green New Deal. Ms. Carol Blackmon of the Southern Rural Black Women's Initiative showed us the power of those who work with the land. Our agricultural economy requires dignified working conditions and family-sustaining wages for all. We must subsidize healthy crops and farm biodiversity instead of monocultures. Our food system must be rooted in principles of community-based agroecology and food sovereignty. Along the way, we must protect and restore our natural habitats, forests, and marshes, which, in addition to being food sources, naturally reduce carbon pollution, and provide a barrier from the storm surges that devastated my community during Katrina.

In our conversations about the Green New Deal for the Gulf South, fisherfolk emphasized that we must defend our last major self-renewing source of food: wild-capture fisheries. The ocean must be protected from pollution, including plastics and petrochemical-based fertilizer and pesticide runoff. The runoff from industrial farms is responsible for the growing dead zone found at the mouth of the Mississippi River, while other harmful algae blooms are decimating our fisheries and coastal fishing communities. The ocean commons belong to us all. A Blue New Deal must end attempts to industrialize the ocean through factory fish farms and large-scale aquaculture. The federal government must limit economic consolidation of fisheries. And they must provide training of historically disadvantaged communities, and economic support for these communities to enter fisheries.

People are already working in the economy of care and repair, and these livelihoods stand to be expanded through a Green New Deal. Affordable housing, access to health care, and quality education are necessary elements of climate resiliency and must be ensured for all.

The Green New Deal must build worker and community power as it is implemented in order to put a check on corporate opposition and enhance the ability of communities to effect change from the bottom up. Green New Deal policy can lead to more community wealth—economic power—by providing pathways for sustainable homeownership, as well as cooperative or public ownership of energy, agricultural, and other firms, large and small.

We can and must increase worker power by overhauling the US labor system through the expansion of collective bargaining rights for all workers, including migrant workers and farmworkers. This begins by repealing pro-corporate labor laws, including right-to-work and at-will policies.

Finally, the complexity of Green New Deal implementation will require government agencies to be more intersectional in nature, with open, better, and more democratic communication and solution-building practices across agencies and with the public. This can create a virtuous cycle of ever more competent administration over time.

The strength of a people-centered design process is that it identifies places where people could slip through the cracks of a one-size-fits-all policy. The challenges facing frontline communities are not rooted in a single issue but rather are the consequence of intersecting breaks and injustices in our social fabric. As we listened, here are some of the considerations we heard.

- Our indigenous community members stressed the need for sovereign rights and said that treaties of all tribal nations must be acknowledged and honored, because this is the foundation of their ability to practice their culture and protect their lands.

- A legacy of incarceration, Jim Crow laws, and slavery has left generations of black southerners with criminal records that bar them from economic opportunity. From the currently and formerly incarcerated, we heard that the Green New Deal must look beyond convictions when designing the workforce of tomorrow.
- Not everybody is able to work, and so despite the Green New Deal's emphasis on good jobs, we must also ensure the safety, dignity, and self-determination of nonworkers—including children, students, elders, people with disabilities, and people between jobs.
- Without fundamental economic rebalancing in response to historic harm, the rights and dignity of black southerners will forever be unrealized. That's why the GND must be understood as a program of reparations for black people, including through land reform. Gulf South for a Green New Deal calls for the codification of the standards for reparations provided in the Vision for Black Lives platform, and we understand the need to draw connections between reparations in the US and climate reparations around the globe.
- Finally, a constant reality of Gulf South life that may be invisible to many outsiders is the centrality of the military to our economy and culture. We must honor the historic significance of the military in Gulf South communities while transitioning investment away from the military and toward job creation in regenerative local economies.

Let's return to where I started, looking at that land loss map on the wall of the community center. Climate migration is a present and growing reality, both within the Gulf South region and from other regions of the country and world. We heard from people who are using a new term—*climate gentrification*—to describe the phenomenon of poor communities that were kept from the waterfront and that are now being priced out of the high ground as the wealthy move inland to avoid sea level rise.

Climate gentrification often happens in the aftermath of climate disaster. When tens of thousands of people left New Orleans indefinitely after Katrina, opportunistic capitalists arrived. Damaged homes were rebuilt sturdier, but at a higher value, and generally outside the reach of poor people who wanted to return home. The price difference is the difference between being able to practice one's human right to return home as a community and being forced to resettle somewhere else—less climate-resilient, and alone.

Rather than seeing displacement and migration as an opportunity for private sector profiteering, what if we saw it as an opportunity to rebuild a social infrastructure, rooted in justice and fairness? We could actually put money into schools and public hospitals and help these institutions prepare for what is to come through climate migration, including the trauma that comes with loss and relocation. We could combat climate gentrification by affirmatively furthering fair, affordable, and equitable housing linked to reliable public transportation. We could protect homeownership by providing material resources to families—especially in communities of color and among others who are vulnerable—for elevating and flood-proofing homes.

We must recognize the right to migrate but also the right to remain, both during acute disasters and during long-term relocation processes. Relocation processes must be self-determined by communities and must assure the social, cultural, and economic requirements for a transitioning community to survive and thrive.

Fifteen years of work in the post-Katrina Gulf South has affirmed my non-negotiable commitments to human rights, human dignity, and human freedom. When we refuse to leave anybody behind, we can't take any shortcuts. Our social, political, and economic systems of extraction must be transformed into systems that regenerate the earth and advance human liberty globally. That requires being strategic—enacting near-term material improvements in our communities in a way that leads to an enduring structural shift toward ecological equity and climate justice.

To survive this next phase of our human existence, we will need to restructure our social and economic systems from root to branch. We must transform from a disposable individual society into a society that grasps our collective long-term humanity—or else we will not make it.

The task is not easy. All of this requires us to recognize a power greater than ourselves, and a life longer than the ones we will live. It requires us to believe in things we cannot see. And it requires us to believe in each other. Let's figure out how to reach a shared liberation together.

Money smelled like mountains of oyster shells rotting in the sun

Genai Lewis

Growing up, I remember eating the same dinner every night—my mother's *moro de habichuelas* and fresh-caught fish from our neighbors.

Thirty miles east of the Apalachicola Bay, nestled deep in the slash pine forest of the Florida Panhandle, is my fishing village, Carrabelle. My dad ran a rural health clinic for our community of roughly one thousand. Most people were poor, without health insurance or money to pay a doctor. My dad treated them anyway, because he believed it was his duty to take care of people, and in turn, they took care of us.

Every other day, visitors would come through our back door to sell fish. Donny and his grandson Gage brought us the catch of the day: redfish, mullet, snapper, shark. Seafood was currency, and in Franklin County money smelled like mountains of oyster shells rotting in the sun. Gage was my classmate, and I remember the first day of kindergarten when he said what so many boys said back then: "I'm dropping out of school when I'm sixteen to be an oysterman with my daddy."

By the time I was eight, my dad couldn't keep his clinic open: too many bills and too few paying patients. For years after my dad closed down, Carrabelle had no doctor. Donny still dropped off fish on the back porch.

• • •

Six years later, the BP-leased oil rig *Deepwater Horizon* exploded in the Gulf of Mexico. Every morning in my high school chemistry class, all we could talk about was the oil: How far had it traveled? Had you seen any oil yourself? There were days when we could smell the oil. Fumes wafted across the waters far out in the gulf and over the barrier islands into our bay. It smelled like melting crayons. We knew that if oil showed up in the Apalachicola Bay, no one would buy our oysters. Terrified politicians opened the oyster season two weeks early, and for a few days the bay was packed with boats that tonged the oyster beds clean. BP set up local offices to distribute payments for people who'd lost jobs and offered their own jobs cleaning up the spill.

Boys like Gage were still around. The ones who had only ever talked about working on the water. With BP's money in hand they bought cars and drove around partying. All night I could hear the screeching of truck tires racing through the backstreets. One morning, there was news of an overdose—partiers realized that they had been driving around with their friend's corpse in the back seat.

Every day at about six o'clock, shifts turn over at the Franklin Correctional Institute, and the evening rush comes through the only two stores in Carrabelle. Most of the kids I grew up with still live there and are either behind the register, stocking shelves, or shopping in their correctional officer's uniform.

These are the jobs that are left, but under the Green New Deal my community could have good jobs on the water again. Jobs like restoring the estuary and seeding oyster beds to heal the Apalachicola Bay. BP's offices are gone, but we are still here, and we can take care of the land and each other if we can win a Green New Deal for the Apalachicola Bay.

NINE

GREEN NEW BINGO HALL

JULIAN BRAVE NOISECAT

In January 2019, I traveled to the Yankton reservation in South Dakota, where, in the Bingo Hall of the Fort Randall Casino, leaders of the Oceti Sakowin, more commonly known as the Sioux, had gathered to reaffirm the International Treaty to Protect the Sacred. The treaty was first signed here in 2013, uniting Indigenous nations in the fight against the Keystone XL pipeline and the expansion of fossil fuel projects emanating out of oil patches like the Tar Sands in Alberta, the Bakken in North Dakota, the San Juan Basin near the Four Corners, and the Permian in Texas and in the pipelines that snake their way across the continent, connecting oil fields to refineries, tankers, power plants, and gas pumps around the world. The year before I visited, the United States rode the wildcatting boom and surpassed Saudi Arabia, becoming the single largest oil producer in the world, pumping more than 10 million barrels per day, according to the International Energy Agency.

The combustion of all that fossil fuel is driving the planet to a new atmospheric chemistry. In 2019, the concentration of carbon dioxide in the air reached 415 parts per million for the first time in 3 million years. The last time that happened, forests grew in Antarctica, the Greenland ice sheet did not yet exist, and sea levels were more than 50 feet higher than they are today. Our species—people we would identify as *Homo sapiens*—did not yet walk the earth.

We did not arrive at this crisis by accident. Every year, the United States provides the fossil fuel industry over $500 billion in direct and indirect subsidies—that's according to an analysis by the radical leftists at the International Monetary Fund. This includes access to cheap leases on public lands, quick and easy permitting processes, and, as we learned at Standing Rock, the cooperation of law enforcement agencies that can serve as mini–private armies.

In the 1970s, scholars coined the term "petrostate" to describe a form of state development built from the revenue of the oil rig. But back when the Saudi Arabias and Venezuelas of the world started drilling, the greenhouse effect was little more than a theory and it would have been inconceivable to power an industrial economy without fossil fuel. Today, of course, we know the science and we have cheap solar panels. We are a $20 trillion economy. We have the knowledge, the technology, and the resources to do things differently. But instead we have doubled down. Since James Hansen testified before Congress in 1988, creating widespread awareness of global warming, we have spewed more carbon into the atmosphere than in all prior human history. Renewable energy is, in many places, now cheaper than fossil fuel, and its price continues to drop. Yet on we go. After three years of decline, US emissions went up 3.4 percent in 2018. President Trump has described global warming as "a hoax" "created by and for the Chinese" and boasts that "the golden era of American energy is now underway." The United States, in league with fossil fuel corporations, has become a uniquely destructive force in global history—one that we may need a new vocabulary to describe. A "necrostate," perhaps.

In the face of destruction, more and more people are rising up. Indigenous communities have been among the boldest of the dissidents, asserting our own authority to stop the construction of pipelines and enact a more just relationship between land and people: to Protect the Sacred.

South Dakota, where we gathered that January, is the land of the Fort Laramie treaties of 1851 and 1868. Each treaty guaranteed the Oceti Sakowin large homelands in the center of the continent. Each was broken even before the ink on the parchment dried. In 1980, the Supreme Court ruled that the United States had violated the 1868 treaty when, in 1874, it opened the Oceti Sakowin's sacred Black Hills to prospectors. The court ordered the government to pay $88 million in restitution. With interest, the settlement is worth more than $1 billion today. The Oceti Sakowin, however, have refused to accept compensation, insisting that their sacred Black Hills are not for sale.

Under the authority of that treaty and others, tribes have been fighting the Keystone XL pipeline, which would contribute 24.3 million metric tons to CO_2 emissions per year—the equivalent of Americans driving more than 60 billion additional miles per year.

The project has been tied up in legal battles for the last decade, though industry experts and anti-pipeline activists think construction could begin again in 2020. In anticipation of opposition, the South Dakota legislature passed an anti-protest law that targeted opponents of the Keystone XL pipeline including ranchers, environmentalists, and tribes with fines for a new crime called "riot boosting." Governor Kristi Noem signed the bill into law less than three days after its introduction. Thankfully, the ACLU successfully challenged the law as an unconstitutional violation of the First Amendment. In the red states that some journalists have taken to calling "Trump Country," the founding documents of this country—the Constitution and the more than three hundred treaties signed with native nations so that these United States could stretch from sea to shining sea—are perhaps the only remaining democratic bulwark against autocracy and climate catastrophe.

These are the circumstances that transformed pipelines like Keystone XL and Dakota Access into national news. For a decade, pipeline politics shifted the center of gravity in the environmental movement. In 2016, more than 1 million people checked in at Standing Rock in solidarity with the fight against Dakota Access. A Puerto Rican bartender from the Bronx was among thousands of millennials who organized in support. Her name was Alexandria Ocasio-Cortez.

Congresswoman Ocasio-Cortez's Green New Deal—like the indigenous movements against Keystone XL and Dakota Access before it—has reshaped the landscape of American politics. One hundred six members of Congress have co-sponsored a Green New Deal resolution, and eighteen presidential candidates endorsed the vision in the Democratic primary, including progressive standard-bearers Bernie Sanders and Elizabeth Warren. Data for Progress, the think tank where I work, consistently finds robust public support for the Green New Deal, which in turn has brought a word not often associated with global warming into the conversation: *justice*.

The congresswoman and Green New Deal advocates, to their credit, have done a great deal to include issues of environmental racism and injustice in the early design of the Green New Deal. In July 2019, for example, California senator Kamala Harris and Representative Ocasio-Cortez introduced the Climate Equity Act, which would prioritize investments in frontline communities in federal climate action. The same summer, New York State passed the Climate Leadership and Community Protection Act, which similarly mandates green investments in low-income communities of color. During CNN's town hall on the climate crisis, nine candidates directly or indirectly referenced "environmental justice."

But one senses that buzzwords like "environmental justice" and "frontline communities" are empty signifiers—empty not because those who invoke them don't genuinely care, but because these terms have yet to be filled with programmatic substance: with what justice might look like for the Oceti Sakowin, for example—or the South Bronx, for that matter. What would it mean to think about communities on the "frontline" of climate change and pollution not only as people in harm's way or as passive beneficiaries of investment but as architects and drivers of a Green New Deal?

So, for a moment, let's turn away from Capitol Hill and the campaign trail and return to the bingo hall of the Fort Randall Casino on the Yankton reservation in South Dakota. What would a Green New Deal look like if we started building here?

Well, as it turns out, the foundation—if not the inspiration—for a Green New Deal already exists in Indigenous communities across the continent. Most, including the best-intentioned, just don't bother to look.

In the values that have called some of the poorest people in the United States to stand up against the largest oil corporations in the world, we might find the courage we need to win this fight—against all the odds.

We might remember that the Constitution protects our right to protest, and that democracy is at least as important and powerful a tool in the fight against climate change as decarbonization is.

In the treaties that the Constitution describes as the "supreme Law of the Land," we might find the basis for a more just and equal relationship between tribes and the federal government: one where pipelines cannot be built through tribal lands without consent.

In the sovereignty of Indigenous nations, we might find new partners for not only a multiracial but also a plurinational democracy. In many places, like South Dakota, where the Oceti Sakowin have stopped the Keystone XL pipeline for the last decade, Indigenous nations have shown how tribes can protect the environment in this perilous age of greed and warming. And through Indigenous nations' inherent and enduring self-determination, we might begin to reimagine and rebuild the state as a home for the many nations of this continent—especially its First Nations.

And from the intergenerational memory and experience of a people who survived a genocide and know what it means to live through the end of the world, humanity might learn a thing or two about how to endure through the climate crisis.

So, what might a Green New Deal built *with* rather than *for* Indigenous peoples look like?

It would look like honoring what came before: the treaties, the tribes, the rivers from which we drink, the air we breathe, the land where we plant and gather our food and to which we return when our time is up. And by finally honoring these things—which have always been there, but which this country has never properly respected or protected—we might build something Green and New.

Back in February at the Fort Randall Casino, I was asked to sign the International Treaty to Protect the Sacred as the lone representative of the St'at'imc Nation in attendance. When it was my turn to speak, I decided to share my great-great-great-grandfather Chief Harry N'Kasusah Peter's song, which many forces—germs, governments, missionaries, and markets—conspired to take away from my family, but which we still remember and sing today.

We may need some new technology to fight climate change—better batteries certainly wouldn't hurt. But, in truth, a Green New Deal can be built mostly through the power of things that have always been there—the sun, the wind, the Indigenous—but whose potential we have yet to fully unlock.

TEN

A WORKERS' GREEN NEW DEAL

MARY KAY HENRY

When Hurricane Harvey swept across their city, wind and rain forced Francisca Reyes's daughter and grandchildren from their home.

They sought shelter with grandmother Reyes, who works as a janitor in Houston. But the storm knocked the roof off Reyes's home, throwing the entire family into chaos.

Seventeen months later, a fire engulfed the town of Paradise, California. That morning, John Allen was at work as a cook at the nearby Enloe Medical Center. A doctor at the hospital warned him that a fire was burning toward Paradise, where John lived.

John's wife was at home recovering from breast cancer, so he raced back to his house to get her out. She survived, but they lost everything, including irreplaceable family photographs and heirlooms.

Tragically, they suffered a much more profound loss that day. John's mother-in-law, who was partially deaf, didn't hear the evacuation alarms. She died when her mobile home burned down.

WORKING FAMILIES' STAKE IN THE FIGHT FOR A LIVABLE CLIMATE

Francisca's and John's hardships are firsthand evidence of how working people—black, brown, Asian Pacific Islander, and white—are often the most vulnerable to damage from extreme weather driven by climate change. Their losses are early indicators of what's at stake as the changing climate starts to reshape the lives of working families.

Francisca and John are members of our union: the Service Employees International Union (SEIU). They are two of over two million SEIU members in jobs throughout the service and caregiving sectors. When climate disasters hit Texas, North Carolina, California, the Philippines, and Puerto Rico, our union family came together to give support and aid to Francisca, John, and other members who suffered devastating losses. SEIU members have also been among the first to respond to these crises in their daily work as health-care workers, social workers, animal control officers, or road and maintenance workers. We've begun to ask what will happen if current levels of carbon emissions cause an inescapable breakdown in the stable climate our civilization has depended on. Janitors and building service workers on the East Coast, who helped clean up in the aftermath of Hurricane Sandy, discussed what would happen to their cities if sea levels rise and storms grow more powerful. Public employees, health-care workers, childcare workers, and other working Californians—who saw so many families devastated by fires—asked what we'll do if the fire season stretches year-round.

Asking these questions showed us clearly that the fight for the well-being of working people can't be separated from the fight to protect the well-being of our planet. We need to join together to fight for more family-sustaining jobs—good, secure jobs that grow thriving communities—and at the same time, we need to fight for clean air and water in our communities.

So we took action. SEIU members mobilized in response to climate injustice. We stood with the Standing Rock Sioux in their fight against efforts to construct a pipeline on their tribal lands, and we joined the Peoples Climate Movement for mass demonstrations in New York. And in June 2019, leaders of SEIU local unions across North America overwhelmingly voted to support the fight to win a Green New Deal. We stepped forward because working people—no matter what our color or where we come from—deserve health, safety, and economic stability. That requires a livable climate, clean air, and clean water.

Climate change is going to change how the economy works, one way or the other. If we don't fight for a just transition, billionaires and corporate special interests will exploit instability to scapegoat immigrants, black and brown families, and union members for problems caused by a warming planet. More wealth will be concentrated into even fewer hands. We have seen what happens when large-scale economic change happens with corporations calling the shots. We live in an era when corporations and economists made sweeping promises that a transition to a globalized economy would create gains for everyone. Instead, entire cities and towns were hollowed out as corporations destroyed middle-class union jobs by shifting work to countries where it's easy to pollute the air and water, and workers have no rights.

For a generation, an influential network of billionaires and corporations coordinated sophisticated legal and political attacks to take away working Americans' ability to organize unions, while simultaneously fighting to preserve dirty energy sources. There's a reason why the union-busters are on the same team as the climate-wreckers. They understand that workers joined together in a union have the power not only to secure jobs and safe working conditions but also to demand clean air, clean water, and a just transition away from fossil fuels. While these powerful special interests claim that we must choose between an equitable, growing economy and a sustainable climate, we know that we need a policy that fulfills both these aims.

The Green New Deal is the sensible action that we need to take to build a thriving country where all working people make progress together. It will keep the cities and towns we inhabit livable, and it will invest in rebuilding communities on the verge of falling behind, revitalizing infrastructure and resurrecting neighborhoods that are getting ignored. It will raise the living standards of working families by creating more good, secure jobs.

As one of the largest organizations of working people in North America, we must link arms with others who share the goal of safeguarding our communities for future generations.

Working Americans want to chart a different course because our economy is not working for most of us. For a generation, real wages for most working people have barely risen. As a result, far fewer working Americans are getting a chance to live their American Dream. A child born in 1950 into a family that made the median household income had an 80 percent shot at earning the same or higher income as their parents by age thirty. But a child born in 1980 into a similar family has just a 50 percent chance of doing better than his or her parents.

Nearly half of America's workforce now earns a paycheck doing service and care work. These jobs are becoming the backbone of the economy, but far too many of them pay wages that are so low that they trap people in poverty, no matter how hard they work. Despite a sustained economic expansion in terms of overall GDP growth, 64 million people in our country are still paid less than $15 per hour for their work. According to the Federal Reserve, 40 percent of Americans don't have enough cash to pay for an unexpected $400 expense like a car repair or a medical bill. Millions of working Americans are paid so little that they qualify for public assistance programs to feed and house their families.

For working families living on the edge of poverty, there is no margin for error. If your workplace closes temporarily because of a storm or fire, that missed paycheck leaves your family choosing between food, rent, and medicine. If a storm or fire damages your home or your car, that emergency can wipe out years of hard-won progress.

In this divided, unequal society, communities of color are hit first and worst by environmental disasters and pollution, due to negligent, discriminatory policymaking, and they are the last to receive aid. Water, air, and soil in neighborhoods where working-class people of color live are more contaminated, causing, among other ailments, higher rates of asthma and respiratory disease in black children. The Flint water crisis is a shameful example. A group of mostly white corporate and political leaders in Michigan treated working-class black families with astonishing disrespect and disregard as they forced them to drink and bathe in poisoned water.

Before, during, and after Hurricane Katrina's landfall in New Orleans, African Americans living in neighborhoods behind inadequate levees to protect from storm surges were abandoned—and even shot at by police officers—as they evacuated. Then they received less support in rebuilding than their white counterparts. There are 96,000 fewer African Americans in New Orleans eleven years after Katrina. The Lower Ninth Ward, a working-class neighborhood, was 98 percent African American before the storm; half of its pre-Katrina population has not returned over a decade later.

Nurses, janitors, and other service workers in Flint, New Orleans, and other cities and towns need policies that prevent future disasters, repair the harms of pollution, and rebuild resilient communities with family-sustaining jobs. The Green New Deal puts racial equity at the heart of this policy vision to make sure that all families benefit, no matter what color our skin is.

WORKING PEOPLE, OUR UNIONS, AND OUR VISION FOR A GREEN NEW DEAL

As much as it is climate policy, the Green New Deal will give working Americans more power to raise their standard of living. It includes policies inspired by the first New Deal, which empowered Americans working in last century's auto, steel, mining, and textile industries to organize unions.

Using the power of collective action, those workers transformed dangerous, poorly paid jobs in factories and mills into family-sustaining work. They fought for a fair return on the profits their work created, raising wages and lifting millions of families into the middle class. By forming unions and staging massive strikes, they also gained voting power that was crucial when it came to winning the broader social reforms of the first New Deal. Through the power of unions, these workers ensured a just transition from a farm to a manufacturing economy.

By 1953, nearly 35 percent of Americans working for wages or a salary belonged to a union. When they negotiated for higher wages, those gains rippled outward to build healthy, thriving communities. But things are different now. Corporations have taken advantage of the shift to a service- and care-based economy to put up barriers that make it harder for working people in today's economy to organize together into unions. Less than 11 percent of working Americans belong to a union. It's particularly difficult for people working in service and care jobs to unite in unions. During the first New Deal, when

Congress passed the National Labor Relations Act, it excluded domestic workers, home care workers, and farmworkers from the right to collectively bargain. Congress kept them out because influential white southern legislators were dead set on stopping the black and brown working people who did most of those jobs from gaining any power.

Over time, corporations have developed new ways to exclude more people. They aggressively redefined the legal definition of employment so they can pretend that they have no legal responsibility for their workers—or any obligation to collectively bargain with us. Corporations have misclassified, subcontracted, outsourced, perma-temped, or franchised millions of jobs so they can evade accountability when it comes to creating jobs that people can actually live on.

Because of this combination of race-based exclusion and job-fissuring, our union estimates that as many as 45 percent of American workers are now effectively denied the opportunity to collectively bargain.

To members of SEIU, the vision for the Green New Deal works in harmony with another set of policies that our members have rallied behind: Unions for All. In the same way that the Green New Deal will force corporations to stop wrecking our shared home, Unions for All will require corporations to sit down with working people and negotiate.

With the Green New Deal and Unions for All policies, we have an opportunity to shift our economy toward family-sustaining work, while not repeating the mistakes of the original New Deal. We can rewrite the rules to empower millions of working Americans to join together in unions, creating a check on the enormous influence that corporations and the wealthy have over our democracy, our economy, and our environment. We can ensure a just transition by maintaining standards of union jobs across sectors, building a bridge of wage support and health-care security for all who need it, guaranteeing that energy sector workers receive training and access to other union jobs, and securing pensions for all workers.

There is precedent for strong Unions for All policies within energy transitions like the Green New Deal. In Germany, workers already have the right to negotiate sector-wide agreements about job standards with employers. Germans are building on that model to gain far-reaching agreements for how to transition away from dirty energy sources. Representatives of German coal producers, coal miners, and affected local governments sat together to negotiate a timetable to wind down the use of coal while providing support for miners and their communities.

We could do the same thing here in key sectors of our economy, but we need Unions for All policies that give working people throughout a region or sector of the economy—not just people who work at one company or worksite—the power to form unions to negotiate with all of the employers in their sector. This would boost standards for people across the sector and prevent companies from trying to undercut each other by reducing pay and benefits.

Working people who are free to join unions will have the power to organize, vote, and mobilize to fight for the Green New Deal and other policies that make life more secure for our families. We will gain the power to join together across difference to stop corporations from shifting the cost of climate change onto working-class communities.

WHEN WE STICK TOGETHER AND FIGHT FOR EACH OTHER, WE WIN

Collective action rooted in social movements is the most direct way to bring about change, and it's how working people will win a Green New Deal.

In November 2012, a courageous group of two hundred fast-food workers in New York City went on strike for a $15 wage floor and the right to form a union, igniting the historic "Fight for $15 and a Union" movement. At first, people ridiculed them because $15 was much higher than any other minimum wage proposal. But they didn't back down. Fast-food workers challenged and then changed the conventional wisdom about service sector jobs in our country, forcing some employers and elected officials to start raising wages.

Other people from all walks of life have stepped up to build new movements: March for Our Lives, Black Lives Matter, the Sunrise Movement, the Women's March, and the Red for Ed public education movement. Now is the time for solidarity.

SEIU members cannot sit on the sidelines while the forces of greed and hate divide us and put the planet we share at risk. Since our endorsement of the Green New Deal, SEIU and youth movements such as the Sunrise Movement and the Future Coalition are already joining forces. We're marching arm in arm in global climate strikes, we're demanding Democratic presidential candidates back a Green New Deal and Unions for All, and we're electing champions of our movement to office.

The land, air, and water bind us together. The Green New Deal creates solidarity and the equity we will need to fight for a livable planet, an economy powered by family-sustaining jobs, and a democracy where every vote counts. We are determined to put more power into the hands of the many so that, united in our vision, we will win.

PART III

ORGANIZING TO WIN THE GREEN NEW DEAL

ELEVEN

PEOPLE POWER AND POLITICAL POWER

VARSHINI PRAKASH

Back in 2007, when I was racing to join my high school's recycling club, I was pretty sure climate change was a big problem, but I didn't have a sense of how far we were from solving it.

The people around me didn't seem too alarmed. After all, a charismatic senator from Illinois was on TV promising to tackle climate change. Less than a year later, Barack Obama was on track to win the Democratic Party's nomination, vowing that together we would halt the crisis: "If we are willing to work for it, and fight for it . . . we will be able to look back and tell our children that this was the moment . . . when the rise of the oceans began to slow."

I *was* ready to fight for it, but all I had was the recycling club. So I dug into the work of saving the world one plastic bottle at a time, while Obama quietly shifted away from the lofty rhetoric of his campaign once in office.

At an off-the-record meeting the White House's "green team" held in 2009, the new administration suggested that environmentalists focus on "clean energy jobs" and avoid climate-focused messaging that would leave the president vulnerable to right-wing critics. "My most vivid memory of that meeting is this idea that you can't talk about climate change," said Jessy Tolkan, who attended the meeting as a representative of the youth-led Energy Action Coalition.

Democrats and DC-based environmental groups then pursued a strategy of compromise. They proposed a cap-and-trade bill—the "American Clean Energy and Security Act"—and partnered with fossil fuel companies to try to pass it in a bipartisan process. The strategy didn't go well. The industry watered down the policy; the Republicans never joined the hoped-for bipartisan process; and the oil-rich Koch brothers financed Tea Party protests against the bill. Meanwhile, climate advocates lacked a large, loud movement to counter the right-wing backlash.

Too many liberals thought the charismatic Democrat in the Oval Office could do it on his own. Van Jones, former Obama administration official and progressive organizer, lamented this lost opportunity: "We thought that by electing Obama, we could just sit back and watch. We went from having a movement to watching a movie." We left climate politics to the politicians, and the politicians were happy to let climate change fall off the national agenda while quietly welcoming more fossil fuels.

Obama told us he'd stop the rising seas, but his administration expanded oil production to historic levels. He gave Shell Oil the green light to drill in Alaska's Beaufort Sea during a period of historic low sea ice in 2012. Three weeks before the BP oil spill, he opened up new offshore oil drilling. On the campaign trail in Cushing, Oklahoma, he boasted: "We're opening up more than 75 percent of our potential oil resources offshore. We've added enough new oil and gas pipeline to encircle the earth, and then some . . ." At the same time, Hillary Clinton's State Department leadership joined with energy companies to promote fracking around the globe.

Following these setbacks, organizers charted a new path forward for Obama's second term that didn't depend on the goodwill of politicians or fossil fuel barons. Climate organizations set their sights on building a movement that could force the climate crisis into national consciousness. The big challenge was how to inspire people without any concrete federal climate legislation on the table. Intrepid grassroots organizers had an answer: get out of DC and fight the industry directly.

By the time the cap-and-trade bill died in July 2010, the Sierra Club's Beyond Coal campaign and local allies had already stopped 132 coal plants, with dozens more on their way to closure. In 2011, a coalition of tribal nations and ranchers led a movement to stop the Keystone XL pipeline, inspiring more communities to directly challenge dirty pipelines and power plants in their own backyards. Steady, relentless organizing against fracking led to bans and moratoriums on shale gas extraction in New York, New Jersey, Maryland, and Vermont. And several groups, most prominently 350.org, saw that the movement to divest from South African apartheid in the 1980s offered a playbook for stigmatizing the industry's catastrophic business model.

Right around that time, a Massachusetts climate organizer put a megaphone in my hands and I found myself looking over the quad, yelling: "The fossil fuel industry's business plan is incompatible with a livable future. If it's wrong to wreck the planet, it's wrong to profit from that wreckage. The University of Massachusetts must divest!"

Our immediate goal was to move university money out of fossil fuel stocks and into more sustainable investments, but the long-term view was that divestment would activate a generation of young people and train them to be organizers. Along the way, we would spotlight the fossil fuel industry's reckless plan to burn five times more carbon than scientists say is safe, and call into question the financial, political, and moral viability of the entire industry.

My education in the divestment movement turned my understanding of the world on its head. I had been taught in high school that change happens slowly, at the whim of rich white men in DC. As my fellow organizers and I looked to the history of social movements for lessons, we saw something different: ordinary people, poor people, black, white, and brown people, women like me, driving social change by banding together to demand freedom, equal rights, and justice.

And these ordinary people could win fast! Although all movement victories rest on decades of quiet groundwork and preparation, the core economic transformations of the New Deal and the major advances of the civil rights movement were each won within a half decade.

This mattered to me as a young activist, because when looking at climate, speed matters. As Bill McKibben often reminds us, we're speeding toward catastrophe so quickly that winning slowly is the same as losing. We need a decisive break from the status quo.

We clearly couldn't assume that Democrats like Obama would join our side after coming into office, so we chose to do what movements in the past had done: build people power and bend the arc of history as fast as possible.

By the end of 2015, over four hundred institutions, including the world's largest sovereign wealth fund, had committed to divest $2.6 trillion from fossil fuels. Peabody Energy, the largest US coal company, cited "divestment efforts affecting the investment community" in its statement of bankruptcy. It seemed like the investments fleeing coal, oil, and gas could mark the beginning of the end of the fossil fuel age. We started to ask ourselves when all this power we had built through divestment would start adding up to the types of big changes in government we knew we needed to stop the crisis.

As it turned out, big political change was right around the corner—just not in the direction we wanted.

In 2016, I watched, surprised and excited, as Bernie Sanders mobilized millions of people into his campaign by talking about the urgent need to solve the climate crisis. Sanders called out the Koch brothers and fossil fuel executives by name and pledged the most ambitious investment in clean energy of any candidate in history. But that's just about all I did—watch. The youth climate movement—bigger than ever thanks to effective organizing around Keystone and divestment—had no coordinated strategy to move our people power to the polls in the most consequential election of our lifetimes.

In the general election between Hillary Clinton and Donald Trump, our movement missed the boat again due to inexperience with electoral operations. When our movement did engage with the elections, it was mostly on the margins. Climate activists peppered Hillary Clinton with demands that she oppose the Keystone XL pipeline (which she did, after many requests), commit to banning fracking (she didn't), and reject fossil fuel donations (she didn't).

Meanwhile, the fossil fuel industry had been taking elections and party politics seriously for decades. They punished politicians who defied their agenda of denial and extraction; they poured hundreds of millions of dollars into congressional lobbying and bought politicians in both parties.

After Trump won, the fossil fuel lobby wasted no time stacking his administration with their cronies. The climate victories of the Obama years were rapidly undone: pulling out of the Paris Agreement, approving the Keystone XL and Dakota Access pipelines, and repealing countless climate and environmental regulations.

I remember a sinking feeling in my stomach as Exxon CEO Rex Tillerson was nominated as Trump's secretary of state. *How did we let this happen?* The truth was, our movement hadn't built nearly enough power. Not only were our grassroots forces well shy of the millions needed to tip the scales, we also lacked allies in the halls of power prepared to champion solutions that actually matched the scale of the crisis. The other side took putting their people into positions of power seriously. We didn't.

The consequences of this approach were clear: nearly all the victories our movement won at the federal level in Obama's second term would be gutted.

If we were going to win a massive, government-led transformation of our entire economy and society away from fossil fuels in the next decade, we needed to build both *people power* and *political power*.

This is Sunrise's *theory of change*, our hypothesis about what we need to do to win: if we keep building people power and political power, we will win a Green New Deal.

Now, if I were reading about this theory of change in 2020, I might think, "Well, duh! What's the big insight here? If you pull the public to your side, mobilize people, and elect friendly politicians, you'll get what you want? That's not rocket science!" You're right: it is intuitive, and that's exactly what we needed.

After Trump's election, I knew there were millions of young Americans who wanted to fight climate change, but I didn't see a youth movement of millions. I knew that if we were going to build a bigger movement—big enough to upend politics and push our country toward historic government action to stop the crisis—we would need to inspire thousands of people to jump into our movement and nimbly move between the worlds of both protest and politics. We'd need to support them through years of obstruction and apparent defeat before finally winning. We'd need to give hundreds of thousands of young people a way to make sense of our movement's progress even when nothing was going our way. A clearly defined theory of change would function as a North Star to keep the movement focused on its ultimate destination. If the theory of change seemed obvious, all the better—that meant it was credible and convincing.

A North Star isn't helpful if it's too dim to see. So, what really is people power and political power? And how do you know when you're on track toward it?

PEOPLE POWER

"We shouldn't have spent so much time in that research committee."

I heard this regret from almost every student divestment organizer I knew. They'd gathered five friends and demanded their university divest. The president had said not now and asked for more data. The students and administration agreed to a special task force to explore the issue. Twelve months later, a long report detailed different options for fossil-free investing. Meanwhile, the campaign had grown no larger than its original five members. No one on campus, never mind beyond campus, had heard anything about the fossil fuel industry's criminal recklessness. No movement had been built. And, of course, the university didn't divest from fossil fuels, because there were only five people demanding it.

It's not an accident that so many activists get sucked into this way of operating. It's what we're taught growing up: focus on the guy in the suit. He's the one who can make it happen. Get all the facts and arguments and persuade him. This idea is so strong in our society that, even in the context of a social movement supposedly dedicated to the "power of the people," we easily forget about the power of the people to make change happen. This is why the North Star—our theory of change—needs to be incredibly clear, incredibly bright, and repeated time and time again so that all the members of the movement know it by heart.

A bright and clear North Star needs to be more specific than "keep organizing protests." If young organizers were going to persevere through loss, we needed an understanding of how to *measure* our people power.

We break people power into two parts, and measure our progress against them:

1. A growing majority of passive support for our cause (demonstrated by polling or other indicators of public opinion)
2. A growing base of active support (demonstrated by leaders and members who take action again and again, through protesting, voting, noncooperating, speaking out, or actively supporting the movement in other ways)

Even if the men in suits say no to our demands—as they will many, many times—we might still be on the path to victory, if we're building enough people power. If we're growing the majority and growing the base, we're winning, because even if our demands aren't met, we can always come back with more people and more support, until we win.

People Power Part 1: A Durable Majority

Movements cannot change laws without winning a majority of passive supporters. Broad public support doesn't guarantee victory, but victory does require broad public support. The major breakthroughs of Obama-era movements happened only when the public embraced their cause.

At the beginning of his tenure, President Obama refused to endorse same-sex marriage. In 2011, when polls found that a majority of Americans backed equal marriage rights for the first time, Obama announced his support. Obama's "evolution" was one of many: between 2010 and 2014, politician after politician changed position, and state after state legalized same-sex marriage not through the courts but through the ballot box. "Gay marriage isn't winning the day because of some singularly persuasive legal argument; it's winning because the battleground has shifted from the court of law to the court of public opinion," wrote journalist Richard Kim in 2013.

There was a similar shift in support for immigration reform preceding President Obama's 2012 decision to grant work permits to young undocumented Americans. Only 32 percent of Americans in 2005 approved of undocumented immigrants holding work permits. After years of young immigrants in organizations such as United We Dream tirelessly working to spotlight the issue, Obama introduced Deferred Action for Childhood Arrivals (DACA) in June 2012; almost seven in ten Americans polled that month approved of the new policy.

Since we're talking about winning a Green New Deal, it's prudent to wonder whether broad support also secured the original New Deal. Mass opinion polls didn't begin until 1936, but there was a litmus test revealing how Americans felt about FDR and the New Deal: the 1934 midterm and 1936 presidential elections. As the president's party nearly always loses seats in midterm elections, it wouldn't have been surprising for Democrats to lose a few in 1934. However, the Democratic Party won an even bigger majority that year, their landslide victory affirming broad public support for FDR's unprecedented project of economic recovery and opening the door to the Second New Deal of 1935—the enfranchising of labor unions via the Wagner Act, the creation of Social Security, and the Works Progress Administration.

A movement's demands may be true and just, but lawmakers rarely step out ahead of public sentiment. Most politicians are not willing to bear the costs of going against the majority of their constituents. That's why Sunrise has made this part of our North Star: building and keeping a durable majority of passive public support for the Green New Deal. With every action we take and every campaign we run, whether we win or lose, we ask ourselves: Is this growing the Green New Deal majority?

People Power Part 2: An Active Base

Now, I wish this were our conclusion: get more than 50 percent of Americans on our side, and we'll win. But that's not the democracy we live in.

In May 2019, 63 percent of Americans agreed that "A Green New Deal to address climate change by investing government money in green jobs and energy-efficient infrastructure" was a good idea. Unfortunately, favorable polling data alone does little to overcome the opposition of the wealthiest corporations in the world. Faced with a choice between broad but quiet public support for a low-priority issue and millions of dollars in campaign donations on the other side, many politicians take the money and ignore the public.

Seventy-five percent of Americans want a wealth tax, yet an army of antitax lobbyists have managed to repeatedly lower taxes for corporations and the rich. Over 65 percent of Americans support bans on assault weapons and high-capacity magazines, yet the NRA has mobilized a relatively small but relentless anti–gun control constituency and campaigned steadily to defeat anti-gun politicians and legislation.

Advocates for climate justice, like most progressives, will always face a financial deficit. A 2018 report from Yale found that between 2000 and 2016, "The fossil fuel industry, transportation companies, and utilities [outspent] environmental groups and the renewable energy industry 10 to 1" on lobbying politicians.

We saw in 2009 how a small, loud, and well-funded faction defeated broad but passive public support when the US tried to pass climate legislation. In the wake of the doomed cap-and-trade bill, political scientist Theda Skocpol autopsied the strategy of the coalition leading the effort, the United States Climate Action Partnership (USCAP). USCAP spent millions on ads to inspire passive public support and sought compromise between environmentalists and corporations to win bipartisan support. But the fossil fuel executives in the coalition double-crossed them, pushing the bill to be "as favorable as possible to their industry," while working with businesses preparing to lobby *against* the bill.

Arguing for the bill on the floor of the Senate, environmentalists realized they had entered a fight for which they were little prepared. They had relied solely on the power of passive public support and their coalition of DC insiders to push the bill through Congress; the bill's opponents not only had the threat and allure of campaign donations but also thousands of industry lobbyists and Tea Partiers mobilizing to stop it. The environmentalists lost, and Skocpol had a clear warning for future campaigns:

> To counter fierce political opposition, reformers will have to build organizational networks across the country, and . . . orchestrate sustained political efforts . . . Big, society-shifting reforms . . . depend on the inspiration and extra oomph that comes from widely ramified organization and broad democratic mobilization . . .
>
> The only way to counter such right-wing elite and popular

forces is to build a broad popular movement to tackle climate change.

Journalist David Roberts echoed Skocpol's warnings:

> National polls tell enviros what they want to hear: In the abstract, majorities always support clean air and clean energy. Enviros mistook these poll results for constituencies. But poll results do not attend town halls or write members of Congress.
>
> Meanwhile, on the right, the Tea Party faction was organized, loud, and incredibly passionate. *An intense, activated constituency beats broad, shallow public support every time* [emphasis mine]. Intense constituencies are levers that move politicians. Polls aren't.

It's clear we need an intense base of supporters to push politicians, but how many people exactly? Political scientists Erica Chenoweth and Maria Stephan provide our best estimate. They looked at every movement from 1900 to 2006, nonviolent and violent, that sought to overthrow a government or liberate their country. They noted the methods, total participation relative to population, and whether the movement succeeded. Their findings are as follows:

Nonviolent campaigns were "twice as likely to succeed outright as violent insurgencies," and "no campaigns failed once they'd achieved the active and sustained participation of just 3.5% of the population." In the United States today, that's a little over 11 million people.

"Active and sustained" are the key words. Their research does not say that a single, 11-million-person march ensured success. It does say that when 3.5 percent of the population engaged *persistently* in nonviolent organizing and action during a movement for liberation—in boycotts, marches, strikes, civil disobedience, and campaigns—the movement was able to overthrow a regime. Consistent mass participation in a nonviolent movement sways the public to ally with the campaign and grind business-as-usual to a halt through noncooperation (more on that soon).

It's not an exact comparison, but the immense difficulties we face in transforming our fossil-fueled economy and political system do share similarities with toppling a dictator. So do we need 11 million Americans? Possibly. We know we need millions, at minimum. To win the Green New Deal, we need a base of people who are ready to knock on doors, strike from work or school, mobilize for a march—again and again. And we need to be constantly growing that base until our numbers overwhelm the opponents of change.

That's why our North Star, in addition to a durable passive majority, is to grow our active base to the scale where millions, eventually 11 million or more, take sustained action for a Green New Deal.

USING AN ACTIVE BASE TO WIN BIG

Let's recap: in order to win, we need a majority of passive support for the Green New Deal and a growing, active base.

Many movements win solely with an active base confronting power holders repeatedly and loudly. In 2009, the right-wing Tea Party movement leveraged an active and loud base to storm congressional town halls and block the agenda of the Obama administration. Indivisible, a grassroots movement launched after Trump's election, successfully saved the Affordable Care Act with similar tactics, relentlessly pressuring their representatives to vote against the GOP's repeal. An active and intense base alone can be enough to win defensive fights.

But our fight isn't only defensive. We can't *block* one bill—we have to *pass* multiple bills in order to rework our entire economy and society in ten years to reduce planet-warming pollution at an unprecedented rate. And our opponents are the wealthiest corporations on the planet.

So we turn for inspiration to movements that won historic reforms despite powerful opposition, beginning with the civil rights movement, which ignited a moral crisis that the nation could not ignore.

IGNITING A MORAL CRISIS

In 2015, 2 percent of Americans ranked global warming as the greatest threat to the nation, placing it outside the top ten priorities. This didn't mean most Americans refused to believe the threat of global warming is real—they just thought the threat was off in the future. When the same group of people were asked, "What do you think will be the most serious problem facing the world in the future if nothing is done to address it?," climate change was at the top of the list.

We need to make the climate crisis, and the Green New Deal, an inescapable, urgent, *immediate*, issue for the broad public and its political representatives. We cannot depend on the worsening storms, fires, and floods alone to push this issue to the top of the national agenda. Our movement must ignite a moral crisis that leaves politicians with no option but to choose a side: the Green New Deal, our generation, and a livable future—or the fossil fuel executives and catastrophe?

To create a national moral crisis, movements make visible the injustices of violence, inequality, and oppression that power holders have long suppressed and perpetuated.

The civil rights movement, in the face of lethal violence and repression, coordinated historic mobilizations and protests that forced racial justice onto the national agenda and ultimately led to the passage of the Civil and Voting Rights Acts. The mobilizations of the Birmingham and Selma campaigns stand apart from other protests due to the sheer scale of nonviolent participation they achieved, their moral clarity of message, and, as a result, their impact on the nation's conscience and politics.

The Birmingham campaign of 1963, demanding the end of segregation and discrimination against black Americans in the city, culminated in 3,500 arrests, overwhelming city jails while 4,000 more marched through the streets. The nonviolent upheaval reverberated across the nation that summer. In one ten-week period more than 750 civil rights demonstrations took place in 186 American cities, leading to 14,733 arrests. Historian Adam Fairclough notes that the prospect of further protests forced the White House to change course:

> For two years [US Attorney General] Robert Kennedy had attempted to deal with each racial crisis on an ad hoc basis. Birmingham finally convinced him that crises would recur with such frequency and magnitude that the federal government, unless it adopted a more radical policy, would be overwhelmed.

Protests against segregation in Birmingham and around the country dramatically elevated the urgency of the issue, raising "the share of Gallup respondents selecting civil rights as the nation's most important problem . . . from 4 percent in early spring to 52 percent by early summer." In a single season, the movement made its demands an urgent national priority.

Addressing the nation on television a month after the Birmingham campaign, President John F. Kennedy echoed the tectonic shift that had occurred in the public's priorities: "The events in

Birmingham and elsewhere have so increased the cries for equality that no city or State or legislative body can prudently choose to ignore them . . . We face, therefore, a moral crisis as a country and as a people . . . It is time to act in the Congress, in your state and local legislative body . . ."

The president promptly sent to Congress a package of civil rights proposals substantially stronger than those he had proposed earlier in the year.

Dr. Martin Luther King Jr., reflecting on these victories in his book *The Trumpet of Conscience*, recalled that legislative solutions to discrimination and disenfranchisement were known and drafted before the Birmingham campaign. It was only when the movement provoked what was irrefutably a crisis that passage through Congress became possible, and then inevitable:

> The Civil Rights Commission had written powerful documents calling for change, calling for the very rights we were demanding. But nobody did anything about the Commission's report. Nothing was done until we acted on these very issues, and demonstrated before the court of world opinion the urgent need for change . . .

It's worth noting that the discipline, courage, and strategic acumen of black leaders in Birmingham had been honed over years of struggle and hundreds of local campaigns. The student-led sit-in movement of 1960 and 1961 alone involved 50,000 participants across the South. Behind every major movement campaign is years of arduous preparation, including many failures. When failures sharpen our message and strategies without dimming our commitment to massive mobilization and moral demands, big change is possible, fast. Dramatic nonviolent conflict can break through the veils of complacency and futility that our opponents use to stifle us, and upend national political priorities within a matter of months.

But we do not need to depend solely on moral suasion to bring decision-makers to our side. Ordinary people have power rooted in the routines of our daily lives. And if we tap that power, we can bring society to a screeching halt to win our demands, just as workers in the 1930s won epoch-defining legislation by striking in such enormous numbers that the federal government had no choice but to enfranchise labor with permanent rights to union representation and collective bargaining.

MASS NONCOOPERATION: THE ULTIMATE WEAPON OF PEOPLE POWER

When I think about challenging the fossil fuel CEOs and their friends, Disney's *A Bug's Life* comes to mind. Hopper, the grasshoppers' gang leader, who controls a colony of exploited ants, worries that a lone bug standing up to him could be a threat to grasshopper rule. To demonstrate, he throws one tiny grain at his henchmen and asks if it hurts. They chuckle, so he throws another. They laugh it off, so Hopper pours out the entire container of seeds, burying the henchmen and imparting a critical lesson: "You let one ant stand up to us, then they all might stand up. Those puny little ants outnumber us a hundred to one. And if they ever figure that out, there goes our way of life!"

Hopper might be a cruel overseer, but he does understand power.

Our entire economy, government, and society depends on us to work, go to school, pay taxes, pick up the trash, and do all the thousands of little things each day that keep the status quo hummin' along. Our cooperation is the bedrock on which everything rests. But as Hopper realized, we ants have tremendous power if we stop cooperating.

Gene Sharp, one of the intellectual grandfathers of nonviolent civil resistance, drew the same conclusion after studying successful liberation movements of the twentieth century: the status quo rests on people going about their business without making a fuss. The course of history changes when ordinary people bring business-as-usual to a halt. Teachers can't educate an empty room, after all.

So-called mass noncooperation is not easy. Organizing massive groups of people to stop participating in society requires more than engaging hearts and minds—it requires wielding coordinated, structural power. This is often the final tactic a movement employs, and it has the highest chance of success.

In the struggle for the New Deal, workers executed the same type of noncooperation as the millions of students now skipping class to demand climate action: the strike. In 1934, a wave of strikes "shook the country" and "panicked" Roosevelt's administration, writes historian William Leuchtenburg. Almost 1.5 million American workers participated in nearly 1,900 strikes that year, including general strikes—shutdowns affecting all industries—in San Francisco and Minneapolis.

From Social Security, to the Works Project Administration, to the Rural Electrification Administration, "every piece of New Deal legislation associated with the high tide of reform," writes historian Steve Fraser, "was indeed the consequence of popular upheaval." The unprecedented uprisings of workers bolstered pro-labor Democrats in Congress, "who now could argue more credibly that only strong measures would restore stability." Congress even passed the Wagner Act, seen by many politicians at the time as strikingly radical, establishing the legal foundation for unions and for workers' rights to collectively bargain, and forcing "employers to accede peacefully to the unionization of their plants," Leuchtenburg writes.

Just as labor strikes during the Great Depression made the Wagner Act possible, so too will strikes of students and workers make the Green New Deal possible. Fifteen-year-old Greta Thunberg began skipping school to protest outside her parliament in August 2018, and just over a year later, school climate strikes ballooned into a global youth movement. On September 20, 2019, 7 million young people around the world walked out of school and work. I walked in the streets of New York, looking at rivers of kids as far as I could see. On stage, I said our generation is starting to take over. I said that this strike is how we stop climate change and win a Green New Deal.

> We gotta be honest with each other if we want to survive. As big as we are, it is still only a fraction of what we need. If we want to win, we are going to need tens of millions of Americans to join us in the streets. If we are going to win, we have to bring society and even our economy to a standstill again and again.

Perhaps Americans aren't yet ready for that kind of mobilization. But back in 2013, before climate change was a top-ranking issue in a presidential primary, a Yale study found that "one in four Americans would support an organization engaging in non-violent civil disobedience against corporate or government activities that make global warming worse (24%), and about one in eight would personally engage in such activities."

One in eight. In 2013. It's time to start organizing.

This is our task: build a movement that not only grows a Green New Deal majority but also sustains an ever-growing, diverse base of leaders and activists that can ignite a moral crisis and can strike until we win. This is the people's power. Let's get to 11 million striking in the streets and see if it works.

POLITICAL POWER

Before Bernie Sanders's rise and Donald Trump's victory in 2016, the movement I came up in saw politicians as weathervanes, twisting and turning with the winds of public opinion. People-powered movements win by changing the way the wind blows, mustering a gale so strong even the rustiest weathervanes pivot.

But as I watched Bernie's campaign turn "radical" ideas into policy proposals that inspired millions and widened the window of political possibility, I was forced to reconsider. Not all politicians are weathervanes; some are champions, using their platforms to change the weather like a movement does.

It's a simple point, but worth spelling out: to pass a bill that decarbonizes the economy and creates millions of good jobs, we need at least one legislator willing to introduce such a bill and tirelessly advocate for it. Then we need a few more politicians who will cosponsor that bill, flesh it out, and twist their colleagues' arms until they vote yes. No grassroots movement is so powerful that it can succeed legislatively without savvy and committed allies in elected office.

That's why our North Star for political power is a critical mass of enthusiastically supportive champions who represent us—not the fossil fuel CEOs—and organize for the Green New Deal in all halls of power, from Congress to statehouses to city councils.

When Sunrise began, we knew we needed to raise the bar for what it meant to be a climate champion. In the 2016 election, no candidate put forth a plan even remotely in line with the scientific reality of the crisis. Hillary Clinton vowed to produce 33 percent of electricity from renewable sources by 2027 and refused to commit to stopping the construction of new fossil fuel infrastructure projects like the Dakota Access Pipeline. Bernie Sanders vowed to ban fracking and named climate change "the single greatest threat facing our planet," but his proposal still had the same long-term goal as Clinton's: reduce carbon emissions 80 percent by 2050. Those plans would mean blowing past the goals agreed to in the 2015 Paris climate accord. Both Clinton's and Sanders's policies would be a death sentence for millions of people.

Yet in 2016, those policies were considered ambitious. Meanwhile, too many environmental groups were (and still are) content to celebrate just about any elected officials willing to say "the science is real." These groups' conception of political power essentially boiled down to electing more Democrats to office, and only rarely criticizing Democrats at all, for fear of losing access to decision-makers.

Unsurprisingly, advocates' low expectations had not propelled Democrats to propose solutions matching the scale of the crisis. Journalist Robinson Meyer, in a 2017 piece headlined "Democrats Are Shockingly Unprepared to Fight Climate Change," noted the gaping hole in the party's approach:

> The Democratic Party does not have a plan to address climate change . . . It does not have a consensus bill on the issue waiting in the wings; it does not have a shared vision for what that bill could look like; and it does not have a guiding slogan—like "Medicare for all"—to express how it wants to stop global warming.

It's easier now to understand what being a champion means, because we have the example of Congresswoman Alexandria

Ocasio-Cortez (AOC). Rather than measuring her policies against the positions of party leadership, she measures them against scientific reality and the moral imperative to protect the vulnerable. And rather than carrying herself with the distant posture of a "sympathetic politician," AOC acts and speaks like a member of grassroots movements.

The evening before our late 2018 sit-in at Nancy Pelosi's office, AOC stood on a table to address a gathering of two hundred young people preparing for action.

> This moment started at Standing Rock. It started by me bearing witness and standing shoulder-to-shoulder [with those] willing to put their bodies and their lives on the line for our future . . . Thank you so much for being a part of this movement and for carrying that torch forward because that is where we come from. We are busting down the doors. Everything is on the line for us.

That's a champion.

AOC joined our sit-in the next day, and three months later she introduced the Green New Deal resolution in Congress. Less than a year later, Green New Deal policies appeared in the platform of every major Democratic presidential candidate. Bernie Sanders vowed to make electricity and transportation 100 percent renewable by 2030. Elizabeth Warren committed to a ten-year mobilization to achieve carbon neutrality across the economy by 2030. Even Joe Biden, who ran a primary campaign distinguished by its sparse focus on policy or ideology commitments, proposed a $1.7 trillion plan to achieve net-zero emissions by 2050, repeatedly noting the inadequacies of the Obama administration's climate plan.

The terms of the climate change debate transformed before our eyes in the course of a year. This wouldn't have been possible without thousands and thousands of young organizers building people power, but it also required a few champions in Congress—Alexandria Ocasio-Cortez, Ed Markey, Rashida Tlaib, and more—who were willing to wield the power of elected office to advance the Green New Deal.

Speaking of Ed Markey, raising the standard for progressive climate policy can also galvanize more establishment Democrats to step up as champions. Senator Ed Markey had long been a congressional leader on climate change, even lending his name to the ill-fated "Waxman-Markey" cap-and-trade legislation in 2009. Since that defeat, Senator Markey didn't gain any traction on climate until the Green New Deal's momentum in late 2018 put new wind in his sails. He enthusiastically adopted the Green New Deal, despite its fundamental departures from the market-driven policies of Waxman-Markey, and soon became the lead Senate sponsor of the Green New Deal resolution.

Aside from using moral protest to change the conversation among the public at large, which opens the door for more Senator Markeys to champion the GND, the main way Sunrise increases our political power is by talking to young people about the Green New Deal and then mobilizing our movement to vote for candidates who would be champions in office. And often, we're not mobilizing alone.

In 2018, Sunrise New York joined a strong community-led coalition in coordinating hundreds of volunteers to knock on thousands of doors to oust the Independent Democratic Conference (IDC), a group of conservative Democrats in the State Senate consistently caucusing with Republicans. This small army of door-knockers helped make the difference in multiple races and succeeded in unseating the IDC. When the new progressive senators took office in 2019, they helped lead the way alongside community organizations in the New York Renews coalition in passing the Climate Leadership and Community Protection Act, one of the strongest state-level climate policies in the country.

In rural Maine, twenty-six-year-old Chloe Maxmin recognized an opportunity to simultaneously address economic challenges and tackle climate change. In a rural district that had never elected a Democrat, she successfully ran an effective campaign for statehouse in 2018. Once in office, she worked with labor and environmental groups—including the Maine AFL-CIO, Sierra Club Maine, and the Sunrise Movement—to build a strong blue-green coalition. In spring 2019, her state-level Green New Deal was the first in the nation to win the endorsement of labor unions. It passed in June of that year.

To be perfectly candid, the electoral contributions of our youth army are not massive compared to our allies in labor unions and long-standing environmental groups; we've got thousands of boots on the ground, but in big races the heavy hitters are posting six- and seven-figure ad buys. Still, every vote counts, and pivotal races often come down to slim margins: two of the IDC challengers Sunrise New York endorsed, Zellnor Myrie and Alessandra Biaggi, won their primaries by just over 4,000 votes; Sunrise volunteers contacted and confirmed more than 4,000 votes for them in each race. Maxmin beat her Republican opponent by only 220 votes. If our work can contribute to a close victory that sends a champion to office, it's worth it.

Working on elections is easy when there's a champion on the ballot. It gets more difficult when none of the options on the ballot are speaking truth or developing Green New Deal–sized plans. Because politics is messy, we published a clear set of political principles to guide our political engagement and keep the movement focused on the North Star of maximizing our political power.

The first political principle is, *We support politicians who will champion the Green New Deal and represent us, not the fossil fuel billionaires.* If a candidate isn't willing to reject money from Big Oil, it tells us a lot about whom they would likely represent. To put them to the test, we joined a campaign promoting the No Fossil Fuel Money (NFFM) Pledge—a commitment "not to take contributions over $200 from oil, gas, and coal industry executives, lobbyists, and PACs and instead prioritize the health of our families, climate, and democracy over fossil fuel industry profits." Over the course of the 2018 elections, 1,300 candidates signed the pledge, and 700 fossil-free candidates were on the ballot come Election Day.

The second principle, *No permanent friends, no permanent enemies*, is designed to prevent the kind of hero worship—or grudge-holding—that keeps activists locked into an outdated stance in relationship to a politician, regardless of that candidate's evolution or backtracking.

The third principle, *We support candidates who represent a significant break with the status quo for their district*, ensures that we're not only working for the most progressive candidates in the most progressive districts in the nation. Instead, this principle encourages our movement to build political power anywhere, so long as there's a candidate who would substantially move things in the right direction.

We needed all three principles to navigate the 2018 gubernatorial election in Michigan. During the Democratic primary, we endorsed Abdul El-Sayed for governor. Abdul was everything we wanted in a champion—a proud No Fossil Fuel Money Pledge signer, unafraid to call for big solutions to the climate that centered racial and economic justice. Sunrise endorsed Abdul when he was polling at around 4 percent. Though he made big gains, Abdul ultimately fell short, winning 30 percent in a three-way race. Sunrise had a tricky choice to make. The Democratic nominee, Gretchen Whitmer, had taken more than $10,000 from executives at the largely fossil fuel utility DTE. Suffice to say, we weren't thrilled about Whitmer.

On the other hand, Michigan was suffering through its eighth year of Republican rule under Governor Rick Snyder, the man who had presided over the poisoning of Flint's children with leaded water. Whitmer campaigned on a promise to fix Flint, as well as to shut down a dangerous oil pipeline running beneath the Great Lakes, threatening 21 percent of the world's surface fresh water. Was Whitmer entirely convincing in her promises? No. Did she represent a dramatic improvement on the despised Snyder? Absolutely. Following our principles, we decided not to make a permanent enemy of Whitmer, who did represent a significant break from the GOP-dominated status quo. Our local Michigan hubs quietly supported her campaign by making this case to thousands of young people, even though we didn't go so far as to call this a full-fledged endorsement.

HARMONIZING PROTEST AND ELECTORAL ORGANIZING

After riding shotgun with our movement's strategy for three-plus years, the thing that strikes me is the relatively harmonious relationship between protest organizing (aimed primarily at building people power) and electoral organizing (aimed primarily at increasing political power).

Our 2018 midterm election get-out-the-vote campaigns not only built political power by electing champions; it also built up our volunteer leadership muscle. After Election Day 2018, we had hundreds of field-tested organizers ready to demand a Green New Deal from the Democrats they'd just helped elect, and that's why hundreds of Sunrisers were able to organize buses to DC a week later for the now-famous Pelosi office sit-in. Our Green New Deal protests throughout 2019 not only built people power by bolstering public support and growing our active base; they also compelled a growing roster of Green New Deal champions to run for office in 2020.

These forms of organizing have flowed well together, though not without friction. In the design stages of Sunrise, we had very few examples of recent movements that had even attempted to work in both the protest and electoral arenas. I hope that my outline of our work in these pages can support future experiments of this type. In "We Shine Bright" (see page 166), Sara Blazevic, Dyanna Jaye, Victoria Fernandez, and Aru Shiney-Ajay explore in greater detail the methods of organizing we've used to create a movement that's prepared for the long marathon of building people power and political power.

Together, we part the sea

Jeremy Ornstein

My great-grandmother gave my nana the core of an apple on her birthday. This story was told at every family gathering, but I didn't understand why the apple mattered so much. Then, when I was seven or eight, my brother grabbed a copy of my grandmother's Holocaust memoirs and went into the bathroom to read the book secretly. When my parents discovered him, they scolded us both, saying "You're too young to read that." I remember thinking, *When will we be old enough?*

A few more years was enough, and one day I followed my brother into the temple auditorium for a presentation on the Holocaust. I remember feeling so proud to be treated like an adult. Then I read that to kill the Jews, the Nazis pretended that the gas chambers were showers. All of my pride fell from me. I whispered, "I didn't know that, I didn't know that." Before I left that room, I had to *grow up*. I learned that there is evil in the world; and I was scared.

Since then, I've learned more: about politicians who sow seeds of hate that blossom into bullets and blood; families torn from home; houses torn by thunder and waves. Stories that haunt—and hurt—all of us, and will for years to come.

But these days, I understand the meaning of Nana's old birthday gift: an apple core, given in the death camps, for a daughter's birthday present! Turns out my history is not only stained by tragedy. When I ask how we will survive the future, I remember that there is also a generosity and love that cannot be taken from me and my people, that cannot be beaten out of our veins.

And this love becomes courage. There's a story my family tells every year at Passover. We're at the Red Sea, and Pharaoh's army is coming—we can see the sharp spears in the sun. Before us, the waves rise and fall, crashing. We are afraid, because we don't know how to escape. Moses sticks his staff in the sand—nothing happens. I am afraid, trembling. The sharp spears already break the skin of those standing at the edge of our huddle. And then Nachson, a kid younger than I am, listens to something within. We watch as he walks into the water. The water goes up to his ankles, waist, then above his head. And in a sudden second we kids follow, running in, and then our elders and our parents follow us, running, crashing into the waves. Together, we bring down something sacred, we *part the sea*, we cross on dry land, we reach the other side. We sing and dance, a little closer to freedom.

No one can see the other side before we part the water. But when you're together, you know it's there. You know you've got to start walking.

TWELVE

WE SHINE BRIGHT

Organizing in Hope and Song

SARA BLAZEVIC, VICTORIA FERNANDEZ,
DYANNA JAYE, AND ARU SHINEY-AJAY

Dyanna

My hands shook as I approached the wooden doors of Representative Nancy Pelosi's office in Washington, DC. Behind me, lined up two by two, were two hundred young people: teenagers, millennials, kids as young as seven or eight. It was November 13, 2018, one week after the midterm elections.

Varshini turned the doorknob, and I felt my heart leap when I heard it click open—*it's unlocked!*

Young people filed into the office of the soon-to-be Speaker of the House, each one carrying an envelope that read *"Dear Democrats, What Is Your Plan?"* The envelopes carried letters and photos of the cherished people and places we might lose to climate change. We dropped our envelopes on Rep. Pelosi's desk and shared what we were fighting to protect.

I went first: *"My name is Dyanna. I'm fighting for my home, threatened by the rising seas in southeast Virginia, and so that no one has to live in fear of losing the places we call home."* I thought about the beaches where I grew up surfing, and the giant pipes installed when I was in high school to suck sand out of the sea and back onto shore to stem the erosion of the coastline.

Forty or so envelopes in, we started to sing:

We are standing for our futures
we are healing what is wrong
we are standing for our futures
and together we are strong.

This song, written by West Virginia organizer Katey Lauer, was our cue to start the sit-in. We linked arms, sat down, and declared our intention to stay.

The next hour was a blur of cameras, unfurling banners, emotions, and song. Rep. Pelosi's office had two large TV screens on either side. CNN was covering the wildfire in Paradise, California, which had already claimed seventy lives at that point. These fires were one of the many reasons we were there—Rep. Pelosi was leading a Democratic Party that had no plan to challenge the climate crisis, and people in her state and all over the country were paying the price.

Suddenly, the twenty cameras outside the office door turned toward the other end of the hallway, and reporters sprinted in that direction. Moments later, Representative-elect Alexandria Ocasio-Cortez peeked in the doorway. She came into the office and gave each of us a high five. Surrounded by shirts that read "12 Years," she called on America to unify behind the Green New Deal.

An hour and a half into our demonstration, I started singing a favorite hype song of the movement, created by Akin Olla, Ilana Lerman, and other friends at the Momentum training institute. The song was written to the tune of Chance the Rapper's "Blessings," and everyone knew that this was the cue to get as loud as we possibly could.

We gonna rise up
Rise up 'til it's won.
When the people rise up,
The powers come down.
They tried to stop us,
But we keep coming back.

Even in our grief at the state of the world, we knew we had to be stronger and louder and bigger to bring down the powers that be, to transform our country. *When the people rise up, the powers come down.* Our voices grew louder with each line as, inside and outside Rep. Pelosi's office, young people clapped and stomped with all the energy we could muster.

Two hours after we'd arrived, the Capitol Police gave us a choice: vacate the office or stay and be arrested. Fifty-one of us refused to leave. The police bound our wrists together behind our backs with plastic zip-ties and led us outside. We sang the entire time.

By the next day, we realized that the power of our protest was far beyond what we had originally imagined. The Green New Deal was the number 1 news story; within twenty-four hours, it had become the subject of thousands of articles. People were signing up for our email list in droves, and hundreds of young people were writing us, wanting to start a Sunrise hub in their town, donate to the movement, or join our next action.

The events of November 2018 happened fast, sweeping us into a whirlwind we'd created but did not entirely control. For our leadership team, this moment and the rapid growth that followed exceeded many of our wildest dreams, but it didn't come as a total surprise.

We had been preparing for this moment for years. The whirlwind was part of a larger vision, a longer plan.

OUR ORGANIZING MODEL

Sunrise Principle #1: We are a movement to stop climate change and create millions of good jobs in the process. We unite to make climate change an urgent priority across America, end the corrupting influence of fossil fuel executives on our politics, and elect leaders who stand up for the health and well-being of all people.

The Sunrise Movement began as a labor of love by a dozen young people, each of us desperate for a strategy to not just halt climate change but also to establish the fossil-free, multiracial democracy we need to weather the storms already on their way. We studied past movements for change and asked, "What can we do today that will alter the course of history?"

Varshini Prakash has detailed our "theory of change," our guiding hypothesis about what we need in order to win a Green New Deal (see page 142). To win, we need:

- People power: a critical mass of engaged people, taking continued action for the movement
- Political power: a critical mass of enthusiastically supportive elected officials who will champion our cause

The Green New Deal movement will need millions of active members in order to overcome the opposition of the fossil fuel elites. Sunrise's aim is to contribute as many people as possible to this overall goal.

For a protest movement to scale into the hundreds of thousands or millions requires stringing together multiple "trigger events" of rapid growth, like the one we experienced following the Pelosi sit-in. This is very difficult, to say the least. Creating a trigger event is 15 percent science, 25 percent art, and 60 percent good fortune. But even more challenging is maintaining unity and focus across the movement *after* a period of rapid growth.

Consider Occupy Wall Street. In autumn 2011, the Occupy movement swept across the country in a mass rebellion against economic inequality and corporate power. At its height, the movement looked near unstoppable, but the momentum papered over serious weaknesses. As it happened, the only things the movement agreed on were a slogan—"We are the 99 percent"—and the tactic of occupying public space. There was no shared goal or long-term strategy. Once police evicted the protest encampments in Lower Manhattan and elsewhere, too many Occupiers found themselves directionless and demoralized, and the movement disintegrated.

Like all decentralized mass movements, Occupy's strength was the *autonomy* it granted participants—anybody could join and take up the brand without having to ask permission. This allowed them to grow fast. However, Occupy struggled to maintain *unity* after rapid growth. As soon as they grew, they fell apart.

Through the Momentum training community, we joined with movement leaders from Dreamers to Occupy to #BlackLivesMatter who had faced this challenge. We asked each other, "How do we build movements that give members the autonomy to support rapid growth, while maintaining unity of strategy and purpose?"

In Momentum we learned to think about movements as living organisms through the metaphor of DNA. In an organism, DNA is replicated as cells divide and the organism grows. The DNA tells each cell how to play its part to help the entire organism thrive. Shared instincts encoded in the DNA of certain species of migratory birds ensure that the birds stay unified in their goal (fly south together!); but having a shared DNA doesn't make every individual bird the same. They experiment and adapt in different ways on their journey south. Similarly, each movement has a DNA of its own—a set of practices and beliefs that guide each member as they make choices about what to do next, as they all move together toward a common goal.

Sunrise's eleven principles, many of which are shared throughout this chapter, are a key tool in maintaining a balance between unity and autonomy. The principles are both guardrails and guidelines, protecting the movement from common pitfalls and destructive behavior while establishing core practices of a healthy movement culture. They remind us of our shared goal ("We are a movement to stop climate change and create millions of good-paying jobs") while encouraging us to embrace experimentation to get there.

Our initial DNA guide took the form of a twelve-page document called the "Sunrise Movement Plan," which has since been updated and adapted into new formats as the movement has grown. Our initial success relied on our ability to successfully give away this DNA to dozens, then hundreds, and then thousands of people, who use it to build Sunrise wherever they live.

To bring the Sunrise Movement Plan off the page and make it a living movement, we built an organizing program with four major components to guide people from their first brush with the movement to active participation, and eventually leadership.

These components are:

- Mass training, which protects the integrity of the movement by training people to implement and defend the movement DNA through in-person and online sessions

- Moral protest, which powers rapid growth by drawing the attention of the media, politicians, and our target membership base of young people
- Local organizing, in which Sunrise members take on the day-to-day work of issue and electoral campaigning, community-building, and slow but deliberate growth through one-on-one recruitment and leadership development
- Distributed organizing, which catches people who encounter the movement online and connects them to volunteer opportunities and nearby hubs

We'll spend the rest of the chapter breaking down each of these components of our organizing model, sharing anecdotes that illustrate how the theories here and in the prior chapter are applied in the real world.

MASS TRAINING: GIVING AWAY THE DNA

Sara

So, how did the movement begin? What did "giving away the DNA" look like in practice?

Training. Training, training, training.

Training is how we grow the movement through disseminating our DNA—our movement strategy, story, and structure—and it is how we develop leaders.

Our first-ever training took place during a sweltering June in Philadelphia. Our founding team crammed into one house and raced to finish turning the movement DNA into slide decks and training modules so that thirty-five of us could learn it. Then, over the next six months, we conducted trainings in ten states. The excitement that came with finally getting to *do* the thing we'd been talking about for a year is muddled with memories of staying up late many nights, eyes straining, to edit slides and rewrite curriculum.

Our early trainings taught us a lot. We learned about all the various things driving the people in the room to join a movement—a longing for more community, a sense of hopelessness in the face of climate chaos, or a desire to take action and fight for something. We honed how we presented our movement DNA and discovered the most effective ways to explain what Sunrise is all about. And we systematized the core concepts and skills that are now the anchors of any training we put on—whether it's online or in person, for ten people in a local training or a thousand people at a regional summit.

These include:

- Strategy: Our theory of change and our four-year roadmap—from 2017 through the next presidential inauguration in 2021
- Story: How to talk about the crisis we face, who is responsible, and what must be done to stop it; how to make our message meaningful by connecting climate change to our lives, our families, and the places we call home, and grounding it in a hopeful vision of a safe and prosperous future.
- Structure and Organizing: How to start a local hub, facilitate meetings, recruit new people and develop them to take on leadership roles, build teams, knock on doors for candidates we support, and take creative local action
- Culture: How to build a healthy group culture and strong relationships, across lines of differences, through storytelling, singing, and anti-oppression practices; culture includes the eleven principles.

We used in-person trainings to launch the movement because we knew that building a movement capable of explosive growth meant maximizing our ability to retain the people who turn out for the first time during "trigger events" and moral protests. We would also have to provide them with the skills to continue organizing once the protests died down. Our early trainings were as much about bringing in new leaders as they were about deepening the skills needed to sustain a movement even when you lose momentum.

Maximizing growth and retention meant we had to be able to train people at scale. That's where "trainings for trainers" came in. Just as we trained newcomers in the Sunrise DNA, we would later invite active hub leaders to participate in a Sunrise Training for Trainers, where they would learn how to put on trainings for their own community. In this way, we designed a mass training program that could grow exponentially, without being bottlenecked by a small core of veteran leaders, fueled by hundreds and then thousands of young people excited to host their first training in their hometown.

By the end of fall 2017, just six months after launching, we had trained 275 people and had 15 trainers. At the date of this writing, in January 2020, we have trained over 4,000 people in person (plus tens of thousands more online) and have over 200 movement trainers. Every month, our numbers go up, and the army of young people grows.

When you ask young leaders in Sunrise about their experience in the movement, they'll talk about a new friendship, an inspiring role model, a sense of purposeful, powerful community—and many of those stories begin with a training. For young people getting their bearings in a world beset by crises, a weekend-long training offers the support and camaraderie of community, a place to see and hear ourselves as leaders even if we'd never thought of ourselves that way before.

MORAL PROTEST

Dyanna

Sunrise Principle #4: We are nonviolent in word and deed. Remaining nonviolent allows us to win the hearts of the public and welcomes the most people to participate. We need maximum participation in order to achieve our goals.

Varshini (see page 150) discussed the need to ignite a moral crisis and build toward mass noncooperation with the status quo. Here, I'll talk about how Sunrisers have used moral protest to grow our movement with these aims in mind. While we're far from the scale or intensity of the Depression-era strikes or the Birmingham campaign, we have been able to shift the national conversation through small and medium-sized actions that prompt a direct confrontation with power holders.

Moral protest transforms the fear and pain of the climate crisis, which we all feel in isolation, into a collective action. In the process, it engages far more people far more quickly than organizers can recruit one-on-one. By giving voice to the unseen suffering and hidden violence of this unjust and unnecessary crisis, moral protest forces the public and the power holders to ask themselves: *Which side are we on?*

YOUNG PEOPLE V. FOSSIL FUEL MONEY

On July 19, 2018, in a town hall in Pennsylvania, eighteen-year-old Sunrise organizer Rose Strauss stood up and took the microphone, looked at climate denier and Pennsylvania gubernatorial candidate Scott Wagner, and asked him a question:

"My name is Rose, I'm eighteen, and I'm really concerned about the future of our country. Two-thirds of Pennsylvanians think that climate change is an issue that needs to be addressed, but you have said that climate change is a result of people's body heat and are refusing to take action on the issue. Does this have anything to do with the two hundred thousand dollars that you have taken from the fossil fuel industry?"

Scott Wagner looked at Rose and responded. "Rose, you know what, I appreciate you being here, and you're eighteen years old, you know, you're a little young and naive. Listen, Rose—are we here to elect a governor or are we here to elect a scientist?"

Rose uploaded the video onto social media and the patronizing exchange went viral, eventually racking up millions of views. All of a sudden young people started sharing their #YoungAndNaive opinions on Twitter, and the *New York Times* called, wanting to talk to Rose.

We seized the opportunity to take action. Sunrisers in Pennsylvania organized a #YoungAndNaive rally and called on others to do the same around the country. In an op-ed in *Teen Vogue*, Rose warned the Scott Wagners of America: "Stand up for our generation, or we'll knock [on] thousands of doors to replace you this November with someone who will."

Rose's challenge to Wagner and our ensuing organizing helped build both political power and people power.

Suddenly, the statewide Democratic Party began campaigning against oil and gas corruption. Democratic governor Tom Wolf's reelection campaign even printed "Young and Naive Voter" T-shirts and stickers. Governor Wolf began to integrate stronger climate messaging into his campaign for the first time.

Reading Rose's story inspired previously disengaged people to get involved with Sunrise. Hundreds of miles away in East Lansing, Michigan, two eighteen-year-olds showed up to an orientation training asking, "Are you the young and naive people?"

Rose's confrontation showed young people across the country how to confront their politicians and helped grow a nationwide campaign to get fossil fuel money out of politics. And the tactic she used, known as "bird-dogging," only requires two people: one person challenges a politician with sharp questions and another records the response for social media and the press. It's a highly visible, highly replicable tactic, and following Rose's bird-dog, thousands of young people challenged Democrats and Republicans in town halls to sign the No Fossil Fuel Money Pledge.

YOUNG PEOPLE V. THE STATUS QUO CONGRESS

A week after the November 2018 sit-in in Pelosi's office, I sat in a room full of Sunrise leaders still flying high off our new success, but challenged by the crucial question: *What do we do now?*

Even though we had put the Green New Deal on the national agenda, we knew we needed to keep upping the ante—what we call *escalation*. Tens of thousands of people were learning about Sunrise for the very first time. We could either rest on our laurels until their attention drifted elsewhere, or immediately invite them into action and turn them into members of the movement.

So we put out the call for bigger action. Less than three weeks after that first sit-in, I found myself running another training in a church packed with over a thousand people. The energy was electric, the building overflowing, and the full balcony covered by a banner that read "Green New Deal Now!" The next day, we arrived at Capitol Hill and lobbied over forty members of Congress before conducting three simultaneous sit-ins challenging Democratic House leadership.

The months of November 2018 through April 2019 felt like holding on to a rocket ship of movement growth and nonviolent escalation, both in DC and in congressional offices around the country. One of the most memorable moments of these whirlwind months happened in the San Francisco office of Senator Dianne Feinstein. Along with our friends in the local group Youth vs. Apocalypse, Sunrisers held a rally of hundreds outside the senator's office demanding she back the Green New Deal. Afterward, around twenty young people, mostly aged eight to fourteen, went upstairs to visit the senator in person. What they experienced was a tour de force of tone-deaf condescension.

"Scientists say we have ten years to turn this around," said one of the young activists. "Well, it's not going to get turned around in ten years," Feinstein replied. "You know what's interesting about this group . . . I've been doing this for thirty years. I know what I'm doing."

The few minutes of heated discussion between Senator Feinstein and the group of children and teenagers went viral within a few hours; #Feinstein became the number 1 trending topic on Twitter, and the confrontation was parodied on *Saturday Night Live* a week later. It was a textbook example of what happens when an establishment Democrat who fails to act on the urgency of the climate crisis is confronted by young people who live with this fear daily. The complacency of an elected leader, previously unremarkable and invisible, suddenly becomes a political liability because of young organizers' moral clarity and nonviolent confrontation.

Office sit-ins and office visits remain essential tactics in our arsenal. They've been used by nearly all of our hubs to grow both nationwide and local campaigns, pushing elected leaders at every level of government to back a Green New Deal.

YOUNG PEOPLE V. THE DNC

Following the spring 2019 sprint to fight for the Green New Deal resolution in Congress, we at Sunrise turned our attention to the presidential race.

We knew that before giving the next president a clear mandate to begin the decade-long transformation of the Green New Deal, we needed to first make climate change a top issue in the 2020 Democratic presidential primary. In 2016, only 1.5 percent of the questions asked in the presidential primary debates mentioned climate change. Because many voters take their cues from candidates and the media, this scarcity reinforced to voters that climate should not be an urgent, campaign-defining priority. We couldn't let that happen again in 2020.

Our strategy was to demand that the Democratic National Committee (DNC) host a "climate debate"—an entire, televised debate that would finally provide the time needed for a deep discussion of the climate crisis and the needed solutions.

In late June, a hundred young people arrived on the steps of the DNC headquarters in DC, ready to sit in inside the building. Our action plan was moot before it even kicked off: we couldn't get past the doors. Locked outside, the group improvised and decided to take over the front steps. It was late June in DC, and it was *hot*. By sunset, the group had dwindled to about thirty of us, red-faced, thirsty, and tired. I felt defeated and powerless, sitting on the brick steps outside the massive political institution that had succeeded in ignoring us for an entire day.

We huddled for a meeting, and someone asked the group, *"Who is willing to stay?"* I braced for everyone to meekly head home. But slowly, one by one, hands went up around the circle. I felt tears welling up in my eyes, moved by the determination I saw around me, and I committed to brave the night, too.

So we stayed. For three hot days and three nights, we sweltered on the hard brick steps outside the DNC. We livestreamed constantly, telling people why we were there and inviting them to join the movement from wherever they lived.

Our DNC protest sparked a nationwide campaign to demand a climate debate. Hundreds of Sunrise hubs pushed their local DNC delegates—people who weren't used to getting any attention from activists—to get behind our demand. Twenty of the twenty-three presidential candidates announced their support. Over two short summer months, the campaign hurtled toward a showdown in San Francisco, where hundreds of Sunrisers crashed the DNC's annual meeting.

In the end, the DNC voted down the proposal, 222–137. We lost the battle with the DNC, but we won the war to ensure climate change became a top issue in the presidential race. The nationwide campaign had made clear that a good chunk of young voters wanted to watch a climate debate, and with no DNC debate planned, CNN seized the opportunity and announced a prime-time Climate Town Hall—seven hours of nonstop climate coverage, featuring one-on-one interviews with every major presidential candidate. By the time CNN ended their town hall, climate change had already received more national coverage than in any other presidential race.

Our three-day holdout at the DNC reminded us how many young people are ready and willing to make personal sacrifices to bring attention to this crisis and to the vision of the Green New Deal. It also taught us to lean on our communities. For every young person willing to spend a night sleeping on the hard brick steps of a DC building demanding justice, there are a dozen others willing to bring food, water, and emotional support to young people sitting in.

ART IN ACTION

Rachel Schragis and Josh Yoder,
Sunrise Movement's Art and Design Leads

In all our different moments of moral protest, there has been a strong visual thread through the art we bring into action. It has allowed our diverse and far-reaching movement to be seen as a united force that must be reckoned with.

If you've seen one image of the Sunrise Movement, it's probably one of these two press photographs from the sit-in at Representative Pelosi's office. The images and signs are powerful, detailed, and disciplined, embodying the strengths of our movement.

In the first photo, we're holding letters we wrote to Rep. Pelosi about what we love and stand to lose due to climate change. They were deeply personal and vulnerable letters, and the sheer number of bright manila envelopes made them a symbol of our collective power and strength. The messaging on the outside, *"Dear Democrats: What Is Your Plan?,"* displayed the unity of our demands.

We always keep our art simple and easy to replicate. When young people started to visit congressional offices across the country, they made their own manila envelopes with *"What Is Your Plan?"* written in black ink on the outside. This demonstrated that we were one big movement, from Raleigh to Little Rock to Tacoma.

NELSON KLEIN, SUNRISE

RACHAEL WARRINER

Studies show that movements that visually appear unified—through using the same chants, songs, and visuals—are more effective at garnering support for their cause.

The visual unity between young Sunrisers across many different geographies is also a source of sustenance and empowerment for

MONA ABHARI AND MICHAELYN MANKEL

our members. It helps people stick around and feel connected to something bigger than themselves. We make art to use at actions, to knock on doors for candidates we support, to set up around the room at hub meetings and community events, to hang on our walls at home and in movement houses. Every time people put on Sunrise shirts, they participate in a story and a project that's bigger than themselves. Art offers a sense of belonging that can be documented

From left to right, hub logos, designer credited: Sunrise Portland, Julian Bossiere; Sunrise Boston, Jamie Garuti; Sunrise LA, Brandon Youndt; Sunrise Cedar Rapids, Jason Snell; Sunrise Durham, Andrew Meeker.

and shared on social media, communicating our values simply and powerfully, and expanding the movement by inviting people to belong with us.

LOCAL ORGANIZING

Aru and Sara

Sunrise Hubs

Sunrise Principle #2: We grow our power through talking to our communities. We talk to our neighbors, families, religious leaders, classmates, and teachers in order to spread our word. Our strength and work are rooted in our local communities, and we are always growing in number.

Moral protests alone do not sustain a movement. To keep people engaged and the movement going between periods of peak activity, we do the slow and steady work of talking to our friends, neighbors, classmates, congregations, and communities and inviting them to join us. Local organizing is all about nourishing the movement by building sturdy structures for members to plug into where they live.

To grow our hubs (our name for Sunrise's local chapters) and develop new leaders, Sunrisers are trained in the core practices of community organizing, including recruitment, relationship-building, forming teams, telling our stories about what inspires us to take action, and local campaigning.

The eleven Sunrise principles provide initial guidance to start a Sunrise hub. Principle #2—*We grow our power through talking to our communities*—reminds us that the best place to start building Sunrise and growing our power as a movement is at home, in our communities. Principle #7—*We take initiative*—empowers all who follow our principles to take action in the name of Sunrise if they recruit others to join them. Principle #8—*We embrace experimentation and we learn together*—makes it clear that people do not need to be seasoned organizers to join, and encourages newcomers to propose and test new ideas for campaigns.

In times of rapid growth, hubs are the structure through which new members can join the movement for the long haul. In early November 2018, Sunrise was powered by about twenty-five hubs. By January 2019, we had around one hundred established hubs. At the time of this writing, in early 2020, we have over 350 active hubs.

Healthy hubs are true communities of friends old and new. Trust and relationships are built through practices like sharing personal stories, singing, sharing skills, and—yes—partying together. Your local hub meeting is where you can go to learn about developments in movement-wide strategy and make local plans. Hubs run campaigns like targeting elected officials to take the No Fossil Fuel Money Pledge and to back the Green New Deal. Hubs endorse and canvass for political candidates; organize protest actions; host fundraisers, parties, and community events; hold grief circles or vigils to respond to local climate-related crises; and show up in solidarity with sister movements in key moments.

SACRIFICE AND COMMITMENT

Sunrise Principle #6: We all have something to offer to the movement. Some of us give time through volunteering anywhere from one to fifty hours per week. Some of us give money. Some of us donate housing or meeting space. We invite our community into the movement by asking for the help we need.

Movements throughout history have been successful because everyday people make sacrifices in their lives to support the movement to reach its goals. Winning the world we deserve requires millions upon millions of people making big and small sacrifices in service of the movement. That sacrifice looks different for everyone—whether it's time spent planning meetings when a student could be finishing homework, energy expended knocking on doors when you could be with your loved ones, or wages lost or classes missed when a person chooses to strike from work or school.

For the 2018 midterm election, we decided to scale up by investing deeply in a core group of leaders who wanted to go all-in. This took the form of a six-month program for seventy-five Sunrise leaders called "Sunrise Semester."

Sunrise Semester participants quit their jobs or took a semester off from school, said goodbye to their friends and families, and moved across the country to build the movement's people and political power. They lived and worked together in "movement houses," building community while supercharging what hubs were already doing in key political locations.

Our inspiration for movement houses stretched back to the supporter housing the Student Nonviolent Coordinating Committee (SNCC) used for its volunteers during 1964's Freedom Summer, an effort to register black voters in the South in the face of violent voter suppression. Contemporary movements have used the movement house model as well; Movimiento Cosecha, a movement to win permanent protection, respect, and dignity for all 11 million undocumented people in the US, had been experimenting with movement houses since 2015.

The seventy-five leaders in Sunrise Semester experienced all the challenges you might expect. We threw a bunch of young people together and trained them to contact voters, recruit volunteers, and organize protests. We learned how to (and how not to) nurture a supportive movement culture, as well as how to have difficult conversations about money, cultivate a spirit of service to the movement, and create shared practices for conflict resolution, and for taking it easy every once in a while.

In total, we contacted over a quarter of a million voters through phone banks, door-knocking, and texting campaigns. But the greatest thing by far that came out of Sunrise Semester was the leaders who grew through the experience.

> I was an angry kid growing up. I was angry at my parents for bringing me into a world that was so painful to live in. I was angry at myself for feeling weak and small in the face of the climate crisis. I was angry at everyone around me because it felt like no one was doing enough to fight for our basic right to live.
>
> I learned a lot of hard skills during Sunrise Semester, like how to knock on a stranger's door and talk about the climate crisis, how to recruit new people, and how to phone-bank strangers. But the most important thing I learned was how to shift my rage at the people around me to the billionaires responsible for the crisis. I started radically loving my community for the first time and I learned to fight for them as hard as I fight for myself. You

> need that kind of love to do this work long-term, to keep you grounded through the ten-hour days, through the moments when you think you're losing.
>
> *Aracely Jimenez, Sunrise Semester volunteer*

At the end of the day, what we asked of participants in Sunrise Semester was a big thing to ask of anyone. And yet it was precisely *because* it was so big that it was so enticing. Young people aren't looking for another petition to sign. They want to do something big enough to solve the problem.

Hub leaders have taken initiative to establish their own versions of movement houses and foster this culture of commitment and community. They've raised money in their communities, petitioned their schools for grant money, and found other creative ways to take time off from school or work so that they can spend as much time as possible in service of the movement. Some have added hours of time volunteering with Sunrise to already busy lives, fueled by their sense of fear and urgency to give their all to this fight. An article about Sunrise in the *New Republic* shares this story from Laís Ramirez Santoro, an eighteen-year-old Sunriser from Pennsylvania who got involved as a junior in high school:

> I go to school, then canvass for four hours after, until it gets dark, then I go home and talk to people about Sunrise and do outreach work for it. It took up a lot of my time. I had a boyfriend at the time, and it did make things tough. It made things tough with my family, it made stuff tough with school because I needed to get good grades. There were so many great things about it, but there was so much I could've done and better balanced if I didn't have that climate anxiety and didn't have that sense of urgency.

Many, many more people have gone all-in in whatever way they can manage—giving an hour to text voters in between multiple jobs, bringing their toddlers with them to protest against their representatives, having difficult conversations with climate-denying friends and family. We are constantly amazed at the variety of ways in which people take the leap of commitment.

BUILDING A MULTIRACIAL MOVEMENT FROM THE GROUND UP

Sunrise Principle #3: We are Americans from all walks of life. We are of many colors and creeds, from the plains, mountains, and coasts. A wealthy few want to divide us, but we value each other in our differences and we are united in a shared fight to make real the promise of a society that works for all of us.

Our fight to stop climate change is grounded in the inherent worth and dignity of every human being. It rejects any political or social force that values any one human being over another. The society we live in is structured and divided by hierarchies: race and ethnicity, wealth, gender, sexuality, religion, nationality, and others. As others have said in these pages, the climate crisis both amplifies and is driven by inequalities that distribute power, rights, and freedoms unjustly along these lines of difference. Many of us in Sunrise were drawn to this movement after we witnessed the ways that environmental degradation and climate change disproportionately affect the most marginalized communities in the United States and abroad—the poor, people of color, women, children, and residents of the global South.

While all these forms of oppression must be challenged, we'll focus here on racism, due to its foundational role in the US and the unique challenges it presents for movement-building. Ian Haney

López (see page 52) describes how winning a Green New Deal—or passing any major progressive agenda—means forging a multiracial movement to tackle racial and economic inequality together. This is no small task. Divide-and-conquer racism has been the strategy of political elites not for decades but for centuries, as Rev. William Barber writes (see page 199). The shores of American history are littered with the wreckage of social movements whose members were unable to establish or maintain cross-racial solidarity.

Addressing racism in our fight for a Green New Deal is both a moral necessity and a strategic imperative. We can't win the Green New Deal without defeating and outmaneuvering the racist dog whistles our opponents use to stigmatize expanded public investments and universal social programs. We can't build a social movement sturdy enough to escalate toward mass strikes without a strong commitment among all members, black, white, and brown, to stand together against racism and all other forms of oppression as we join together in nonviolent actions that can pose unequal and unjust risks for working-class people and people of color.

So how exactly do we do this? We stand against racism through the messages we amplify in the public arena, through the alliances we build and the candidates we endorse. We also strive to practice anti-racism within our movement. To participate in Sunrise is to accept the challenge of building a multiracial, cross-class movement: to commit to learn, be open to your own transformation, and to relentlessly face down the reality of racism.

This journey in Sunrise has been ongoing and imperfect. Following the guidance in Principle #8 to *"learn through honest mistakes followed by honest conversations,"* we've made a lot of mistakes and had a lot of conversations, often prompted by leaders of color agitating lovingly to make the movement stronger. We'll keep learning and experimenting together as we go. Here are a few things that we've done.

Anti-oppression education in universities, workplaces, and social movements is often highly intellectual and focused on understanding the facts of the matter. It can be very effective for some—but the approach by itself often falls short of helping us foster new relationships across lines of difference. The dominant culture of isolation and individualism, compounded by the still-present physical segregation of our society, prevents many people from being able to form meaningful and caring relationships across racial divides. Sunrise trainings and hubs use a few key practices to create opportunities for our members to build such relationships.

One tool we use widely is "storytelling and resonating," a form of personal storytelling and active listening. Telling stories often helps illuminate the ways that our differences shape our lives, and even more often reinforces the commonalities in the joys and suffering that make us human.

Singing is another important connective practice. Songs unify us in ways that can feel mysterious, magical, and sometimes even spiritual. Songs help us stay calm and resolute during a demonstration; channel anger, sadness, or grief; raise each other's spirits; and share history, heritage, and culture with one another.

These practices create the opportunity for relationships across difference, which is absolutely essential, but still not enough. As new members join the movement on a daily basis, their oppressive behaviors (we all have them!) join the movement with them. Members new and old have a responsibility to engage in ongoing work to combat racism and other oppression within their own attitudes and in society at large.

One way we support this work is through trainings for members ready to deepen their leadership through an anti-racist lens. In late 2018, Sunrise leaders put on our first Leaders of Color training, which focused on investing in and building a deeper community among leaders of color in the movement. This training engaged participants in healing from the generational trauma of racism, and in supporting one another in the shared challenges of being a leader in majority-white hubs. The movement now holds these leadership trainings for thirty to forty Sunrise organizers every few months.

We also have an Anti-Racist Training for White Leaders, where white folks work to see themselves as collaborators and leaders in the project of combating racism. Like the Leaders of Color training, our trainings for white leaders combine guidance in processing experiences as white people with tools to address harm and guide healthy group culture.

The only way to win a Green New Deal is to build a massive, multiracial, and cross-class movement that engages millions of people. Doing so requires confronting racial and economic oppression, both inside our movement *and* in the narrative and strategic battles we wage in public.

DISTRIBUTED ORGANIZING

Victoria

Sunrise Principle #7: We take initiative. Any group of three people can take action in the name of Sunrise. We ask for advice—not permission—from each other to make this happen. To make decisions, we ask ourselves, "Does this bring us closer to our goal?" If yes, we simply do the work that is exciting and makes sense.

In the midst of the chaos of our sit-in in Pelosi's office, with hundreds of young people united in song and the Capitol Police giving warnings to activists in the hallways, I found myself at the other end of the hall, kneeling on the ground with my laptop propped up on a recycling bin, computer connected to phone connected to battery pack. My mission: to invite the thousands and thousands of people watching our protest online into active and ongoing roles in the movement.

In moments of sharp moral protest, onlookers often feel inspired to take action with the movement from wherever they are. As people tuned in to the livestream of the young people in Pelosi's office, a team of seven online volunteers had only *minutes* to catch curious and inspired onlookers. We messaged the thousands of livestream watchers, inviting them to a video conference call the next night to learn how to be part of the movement. The goal was that people would soon become part of the *active base*, graduating from "liking" and commenting on their support for a Green New Deal to showing up in person in support of it.

As footage of our sit-in went viral, over a thousand people signed up to join the call. We ran the call as though it were an online rally: Sunrisers who recently got involved told powerful stories about the moment they chose to join, and the fear and power they felt when they took action. We told participants that the most important thing they could do *right now* to push Nancy Pelosi and members of Congress to back the Green New Deal was to visit their representatives' offices wherever they live. We shared a Google doc with sample chants, scripts, and step-by-step guidance to plan a local office visit.

People were itching to do something more than retweet, and we knew that if we asked them to do something big and inspiring, and gave them support to be successful, they might just do it. That's how, only one week after two hundred young people occupied Pelosi's office, young people brought the demand of the Green New Deal to two hundred congressional offices across the country.

This is the work of *distributed organizing*: growing the movement by reaching thousands of unengaged people online and plugging them into on-the-ground activities. Everyone who sees Sunrise actions online, who likes a post on social media or clicks a link in an email, is asked by a volunteer team to get involved. The invitation is sent via direct message or text message. We support people to go from online supporter to in-person action as quickly as possible.

The "three-person rule" in Sunrise Principle #7 is critical to new people entering the movement and being empowered to take action. Building a movement to scale means letting go of the desire to control everything, trusting people who feel the call of leadership, and allowing the principles and DNA of Sunrise to guide our shared work.

We've found that when people are encouraged to take initiative, they take on greater risk and leadership than they would have ever taken if they were waiting for someone else's permission. Paul Campion from Sunrise Chicago shares the story of how he got involved in Sunrise:

> I had been on the email list for over a year, following along on social media but not doing anything actively. I got the email to go to the November sit-in right after the 2018 midterms. I had an exam so I couldn't go, but I watched the livestream. I remember feeling like, *Holy smokes! Whatever my other priorities are, they aren't as important as this!* Seeing the power demonstrated by those young people—I hadn't seen that before. And the hope they had—I hadn't felt that before.
>
> When it came time for the next action, on December 10, I was determined to get myself there. I got five friends from Chicago, my cousin from South Bend, and a friend from Minnesota to all go to DC with me. It was the first time I had risked arrest, and between being in the training with hundreds of people and taking the action the next day, I remember wondering, *Why has it taken me so long to find this place? This is so much of who I want*

> *to be.* I met a couple people there who had Chicago connections, and we all went back over the winter holidays. At first I didn't want to start a hub—I didn't feel like I was the right person to do it. But when I realized it might not happen unless I did it, I invited those people I'd met over, and together we planned the launch meeting and got things going.
>
> Sunrise is about telling young people that we can do whatever we set our minds to. We don't have to wait till we know more; we can try, and mess up, and try again. None of us has ever done this before—we're all figuring it out. And even just you trying is a huge step in the right direction.

For people who aren't inclined or able to take on an in-person role like Paul, we design additional roles that people can fulfill online. Hundreds of people in dozens of volunteer teams support critical functions of Sunrise. A few examples: the data entry team, which enters and manages data to support hubs to reach the people they are organizing quickly and efficiently; the texting and phone-banking teams, which communicate key information across the movement; and the inbox team, which makes sure organizers in the movement get their questions answered and get the support they need in order to do their jobs.

Sunrise has always been rooted in the belief that transformative change comes from the power of everyday people doing what they can from where they are. No organization or network of organizations can ever pay all the people we need to make a Green New Deal the road map for our country, and it's the contributions of volunteers that sustain the movement and power it to be stronger each day.

Distributed organizing taps into a courage that says, *Maybe I can do something—maybe I can call my representative or text five friends or host a house party—and maybe, with a few of my friends, we can participate in a movement to shape the course of history.*

This little spark of courage is what we activated in ourselves when we started Sunrise, and it's what we encourage and invite from everyone who makes the movement their own today. Everyone is welcome, and anyone can make change. If you share the goal and vision of the Sunrise Movement in your heart and you follow the principles of the movement, you can act in the name of the movement and claim the movement as your own as much as anyone who has been in it before you.

Sunrise Principle #11: We shine bright. There are hard and sad days, to be sure. This isn't easy work. But we strive to bring a spirit of positivity and hope to everything we do. Changing the world is a fulfilling and joyful process, and we let that show.

Blue skies in America

Saya Ameli Hajebi

My name is Saya Ameli Hajebi. I am eighteen years old and I grew up in Tehran, the capital of Iran, one of the most polluted cities in the world.

When I was little, I never needed to read air quality reports in the newspaper. I knew pollution was bad when I saw my brother straining to breathe and his skin breaking out in rashes.

I didn't believe my mom when she reminisced about the white cotton clouds dotting the blue skies of her childhood. If you had asked me what color the sky was, I would have told you with conviction it was gray.

One day my mom took my brother and me on a hike. When we reached the summit I looked out over the thick gray canvas that covered my city. Above me, it looked as if someone had melted my bright blue crayons into a clear sky. I felt cool, fresh air fill my lungs and I turned around to see my brother's chest rising and falling with ease. I wanted those blue skies to surround him forever.

Soon after, we left Iran for the US, saying goodbye to my grandparents and the laughter that rang out every time my family shared a meal of saffron rice and *bademjan.* We left a city where smog canceled school, where protesters risked their lives for a chance at justice and were met with violence. We left in search of freedom, democracy, and blue skies.

I have now lived in Boston, Massachusetts, for nine years. During my first Sunrise meeting, I learned that Robert DeLeo, the Democratic speaker of the Massachusetts House, had accepted over $40,000 in campaign donations from the fossil fuel industry, and in 2018 his State House watered down an ambitious, comprehensive climate bill our State Senate had passed, producing a compromised bill that was nowhere near what our state and our planet needed. Was threatening my brother's health only worth a few more dollars in Robert DeLeo's campaign coffers? I felt angry, but when Varshini Prakash asked if anyone was willing to stand up and speak at Robert DeLeo's office that day, I hesitated.

I remembered a cool autumn night in elementary school. I was peeking out from behind my bedroom door in Tehran, watching my mother crying. Her voice broke as she explained to my father how she had been arrested for showing too much hair under her hijab. No matter how many times I tried to hug and comfort her, I couldn't give back what she'd lost that day.

Varshini repeated her question. Who could speak for our movement at the statehouse? I had learned early in life that people who tried to create change would be arrested, humiliated, and sometimes even hanged in Iran . . . but the image of my brother surrounded by clear skies kept tugging at my sleeve, willing me to raise my hand. I am tired of feeling scared that my brother will struggle to breathe again because the people in power refuse to take action.

The next day, I stood under the statehouse dome surrounded by Sunrisers and said: "Speaker DeLeo, you will either stand with my generation and protect our communities, or we will replace you with someone who will."

Speaker DeLeo didn't sign the No Fossil Fuel Money Pledge that day, but we returned to the statehouse again and again, growing from tens to hundreds to thousands in just a few months, until Speaker DeLeo passed a $1.3 billion climate bill. It's not enough, but now we know we have the power to change politics, and we won't stop until we win a Green New Deal. We don't have to fear gray skies anymore.

THIRTEEN

A THIRD RECONSTRUCTION FOR OUR COMMON HOME

REV. WILLIAM J. BARBER II

Along the Mississippi River, between Baton Rouge and New Orleans, Louisiana, land where black people were once enslaved on plantations is now being poisoned by petrochemical plants that have given the place a new name: Cancer Alley. In St. James Parish, I have visited with families in which every member has some form of cancer. At the Mt. Calvary Baptist Church in St. John the Baptist Parish, a local resident named Robert Taylor told neighbors in the fall of 2019 about the toll of watching his family and neighbors die. Next door to an elementary school in his predominantly black community, toxins measure 370–700 times the levels that are considered safe. Robert's daughter has a rare disease that her doctor told her she had a one-in-five-million chance of contracting. She has since learned that three other neighbors are dying of the same disease.

While Cancer Alley is a glaring example of environmental racism, it is also a microcosm of the death-dealing legacy that America's plantation economy has wrought on the whole planet. Built on stolen labor and stolen land, this system produced unprecedented wealth that allowed America to become a world superpower. But it also established institutional habits that normalized destruction. From the American colonists' earliest encounters with the Indigenous people of this land—what they call Turtle Island—Native wisdom has warned against the unsustainable practices of exploitation and extraction. "Indigenous peoples have witnessed continual ecosystem and species collapse since the early days of colonial occupation," Leanne Betasamosake Simpson, an activist and scholar from the Nishnaabeg nation, told *Pacific Standard* magazine. "We should be thinking of climate change as part of a much longer series of ecological catastrophes caused by colonialism and accumulation-based society." Indeed, our First Nation sisters and brothers help us see that we cannot save the planet—what Pope Francis rightly calls our "common home"—without first reconstructing the systems that have overlooked and marginalized non-white people from their inception. In short, there is no Green New Deal without a Third Reconstruction that learns from Reconstruction history the kinds of organizing that have been essential to each stride toward a more perfect union.

We cannot separate the question of whether we can survive together on a warming planet from the question of whether we can redeem the promise of liberty and justice for all in this nation. We are, in the powerful image from the biblical story of Noah's ark, all in the same boat. A future free from the polluting economy of extraction is inextricably tied to a future free from the systemic exploitation of white supremacy, and this future requires reconstruction and transformation. When we are willing to look honestly at our history to understand the foundations of interlocking injustices, we can see in that same history examples of multiracial movements that have worked to reconstruct America from its very beginnings. Indeed, no expansion of democracy that we celebrate today would have been possible without the fusion organizing that brought people together in moral movements to work for abolition, just labor practices, women's suffrage, civil rights, and LGBTQ+ rights. So in order to look forward to a Third Reconstruction, we need to take time to look back and remember how the systems we have inherited came to be.

THE BIRTH OF A NATION

Four hundred years ago, a Dutch warship, the *White Lion*, sailed from the Netherlands—under a Dutch flag but with an English captain and crew. Aboard the ship were privateers licensed by their governments to plunder Spanish vessels. They robbed a Portuguese slaver of roughly sixty of its human cargo. The privateers sailed the twice-stolen Africans to the English colony at Jamestown, Virginia. There, reported local leader John Rolphe, the *White Lion* traded the settlers "20 and odd" Africans in exchange for food and provisions.

The Africans who disembarked at Jamestown landed among colonists who had no cause to regard themselves as "white." The Africans who arrived on the *White Lion* may not have been ordinary indentured servants, and they certainly did not fold neatly into an existing institution. Most Europeans in Virginia arrived as indentured servants, bound for seven years, after which the colony granted them a plot of land. If the legal status of the Africans as they clambered from the ship that day remained uncertain, their color was not laden with the meanings of "race" that slavery would inscribe upon the bodies of their descendants.

In Jamestown in the early seventeenth century, unfree labor was both black and white, and had yet to be cleaved by color and clotted into "the peculiar institution" of slavery. But the racial categories that we now take for granted were soon born in this place and for this specific purpose: to justify plantation capitalism and prevent poor black and white people whose labor sustained it from uniting in a fusion coalition that would expand the promises of democracy to them and their families.

The greatest threat to the consolidated power of colonial Virginia was Bacon's Rebellion, an uprising of disenfranchised descendants of Europe and Africa who banded together to overthrow the plantation owner plutocracy. It took the British navy to put Bacon's Rebellion down. To prevent such an uprising again, the Virginia colony set about formalizing racial categories and a racial hierarchy.

Legal structures around color and status changed sharply. The Virginia Slave Code of 1705 fully consolidated the system of racial and hereditary bondage. Bad biology, sick sociology, political pathology, evil economics, military madness, and heretical ontology were written into law.

To justify the unfolding robbery of racialized and hereditary slavery, the owning class established the doctrine of white supremacy. White supremacy became both pillar and signpost of these evil arrangements; it grew into both a political program devoted to white dominion and an unconscious array of assumptions of inferiority and superiority that we carry in our heads, whatever their hue. By separating us into inferior and superior, the planter class reinforced their ultimate superiority. The plantation system told the white man that he might be poor, but at least he was better than a black man.

When we do not know this history, we easily fall prey to racism's lie that insists an exploitative and extractive economy is as normal as the racial categories that were developed to divide people. But we also forget that those categories were developed to blunt the power of fusion movements. From the very beginning of the American story, an extractive economy that exploited labor and land was challenged by the power of a fusion coalition that could unite exploited people to demand that the promises of a representational democracy extend to all people equally.

Americans tend to believe that "race" is something real, an undeniable assignment by the natural world. "Racism," we assume, stems from this bleak but biological condition. Ta-Nehisi Coates explains that this delusion renders white supremacy "the innocent daughter of Mother Nature." This confusion leads many to bemoan the death machine of the Atlantic slave trade and lament the Trail of Tears in the same way that we mourn a tornado—without a twinge of remorse. But, Coates reminds us, "race is the child of racism, not the father."

The father of racism is the need to justify the brutal and hereditary bondage of chattel slavery. The reason was staple crop agriculture—tobacco, rice, sugar, coffee, and cotton. Enslaved labor was the oil of the eighteenth and nineteenth centuries—a necessary requirement to the false promise of eternal economic growth that still exists today. We cannot free ourselves from our dependence on oil without freeing ourselves from the lies that were used to justify our dependence on stolen land and stolen labor.

Another pernicious lie of white supremacy is that it is to the benefit of white people. In truth, white supremacy is as poisonous to white people as it is to people of color—and nothing highlights this fact more clearly than the climate crisis.

As the Indigenous people of this land have long known and our scientists have more recently confirmed, an extractive and exploitative capitalism can transform the blue-green jewel we inhabit into a cold and empty stone spinning in space. When the waters rise, the fires burn, or the droughts come, the poor always suffer first. But they are told the same racist lie—that the real problem is the brown immigrant, the lazy black mother, or the Muslim neighbor.

By continually pitting poor and marginalized people against one another, racism has allowed a political party in the United States to maintain power even as it imperils life on this planet.

White supremacy had a beginning and it will have an end, whether in the eternal silence of an uninhabitable earth or on a planet far more just, loving, free, and peaceful than the one we are trying to save. The judgments of God are just, but they are not always pretty. Our long-delayed day of reckoning has come, and, truth be told, we are both the accused and the jury. We must decide our own fate.

REVIVING THE HEART OF AMERICA

When we are honest about our past, we can see how the idea of liberty has progressed alongside the racist ideas that were used to justify the murderous sin of chattel slavery and our abuses of the earth. Democratic dreams and systemic racism are both our legacy; heroism and hypocrisy our tangled birthright. Visions of universal love and heresies of slaveholder religion walk hand in hand throughout our history. It is increasingly clear, however, that neither democracy nor our common home can endure if we cannot dismantle systemic racism. The paddles lay back on the pier, the canoe lunges toward the falls, all of our children are in this boat, and all we can manage to do much of the time is bicker about who left the rope untied. All that most of us know for sure is that none of it was our fault.

But, as James Baldwin said, "we made the world we're living in and we have to make it over again." The legacy of America's fledgling democracy, the question of whether we have anything more than a trillion tons of plastic to offer the world—all these matters are bound up in our capacity to change.

While the patterns and practices of the plantation economy have plagued our common life, we have also seen powerful movements toward a more perfect union in our shared past. African Americans have joined hands with whites in the North and the South, remembering the potential of the fusion coalition that first shook the plantation system in Bacon's Rebellion. Within four years of the Civil War's end, white and black alliances controlled every statehouse in the South. Together, they elected new leaders. Almost all of the southern legislatures were controlled by either a predominantly black alliance or a strong interracial fusion coalition. They hammered out new constitutions from a deeply moral perspective.

What was the agenda of this Reconstruction effort? In order to expand democracy, they knew they had to both expand voting rights and build coalitions across former lines of division. Black Republicans united with white Populists to work together, even though many poor white farmers in the Populist Party still harbored racist ideas. These coalitions built the first public schools in the South and granted all persons a right to public education in new state constitutions. The constitution of North Carolina stated that "beneficent provision for the poor, the unfortunate, and the orphan is one of the first duties of a civilized and a Christian state." Labor rights and the right to "enjoyment of the fruit of your own labor" were included in 1868, long before the Knights of Labor arrived with their first southern campaign. They knew then—black and white together—from a moral fusion perspective, that labor without a living wage is just a different form of slavery.

But in four years, the experiment of the First Reconstruction faced powerful and immoral opposition. And we must understand that opposition then to understand America right now. Many former Confederates saw black citizenship and interracial alliances—fusion coalitions—as inherently illegitimate. They created the Ku Klux Klan to terrorize white fusionists whom they viewed as race traitors. They attacked black leaders. Conservatives began to rail against taxes and attack government programs, rendering state governments unable to lift up former slaves. They used old racist ideas to instill fear in white Populists and attacked fusion coalitions that sought to extend the rights and opportunities of democracy beyond a privileged few.

Why were they doing all of this—rolling back voting rights, taking away criminal justice reform, and undoing equal protection under the law? They said they wanted to "take back America." They said, "We came to redeem America." They distorted this nation's moral narrative to justify immoral activity. And by the turn of the century, many of the gains of the First Reconstruction had been overturned.

This same pattern of progress and backlash repeated itself during America's Second Reconstruction—what we often remember as the civil rights movement. When black and white joined together in the Student Nonviolent Coordinating Committee, for example, they challenged Jim Crow segregation. But they also organized the Freedom Summer in Mississippi and worked with other civil rights organizations in the Selma campaign to win the Voting Rights Act. As in the First Reconstruction, the expansion of voting rights was central to the Second Reconstruction, and the expanded franchise led to real political power for working people. Black, white, and brown people formed a coalition that transformed immigration policy, increased access to housing, guaranteed environmental protections, and waged a war on poverty. By the late 1960s, dozens of organizations representing African Americans, poor whites, Chicanos from the Southwest, and Native Americans had united in a Poor People's Campaign led by Rev. Martin Luther King Jr. and many others.

Beyond the typical frame of left versus right, Republicans versus Democrats, fusion politics made it possible to work together for moral issues in the public square. We must never forget that a Republican administration initiated the Environmental Protection

Agency because an environmental movement emerged from the Second Reconstruction to insist that everyone in public life take seriously our moral obligation to care for the planet.

Of course, America's Second Reconstruction was also met with an extreme backlash. Political operatives of the New Right developed the "Southern strategy," using racist dog whistles to play on the fears of white voters and consolidate political power. Corporate interests invested in campaigns to delegitimize government, gut unions, divide fusion coalitions with wedge issues, and deny climate science. These regressive forces had to exert extreme pressure to maintain control precisely because the potential power of black and brown voters forming a fusion alliance with progressive whites presented an enormous threat to the status quo.

The political pathology that undergirded slavery and formed the backlash of the post-Reconstruction periods is still alive in the constant assault on democracy we are witnessing today. Since 2010, twenty-three states have passed racist voter suppression laws, including racist gerrymandering and redistricting laws that make it harder to register; reduced early voting days and hours; purged voter rolls; and passed more restrictive voter ID laws. In the midst of this, Mitch McConnell and his enablers in Congress have refused to restore the Voting Rights Act, since the Supreme Court gutted Section 5 in 2013. November 2016 was the first presidential election in fifty years without the full protections of the Voting Rights Act. If we call Strom Thurmond a segregationist because he filibustered the 1957 Civil Rights Act for over twenty-four hours, what do we call an entire political party that has conspired to block the restoration of voting rights for more than six years?

We must begin to think of our present crisis and present task in terms of a Third Reconstruction—an expansion of American democracy to include poor people of every race, creed, and culture, and a strengthening of the social contract to ensure fundamental economic and social rights for all. The project of a Green New Deal must be in lockstep with a Third Reconstruction to revive the heart of America's democracy.

The Koch brothers and Donald Trump are only the latest expressions of the unholy alliance between capital and white supremacy that has shaped this country since before its founding and brought us to our present peril. As long as hundreds of millions are spent to hijack election results and install racist, antidemocratic, climate-change-denying candidates to all levels of government; as long as elected officials are the mouthpiece and vessel for profit-driven corporations; and as long as young people, people of color, and poor people are systematically excluded from having a voice in our politics, we cannot win the transformative socioeconomic change we need to stop climate change.

We cannot understand how our crises of climate and democracy intersect without also understanding the moral crisis of poverty in America. We know that 140 million people in the United States alone are already poor or at extreme risk of being poor. When you add the impact of the climate crisis, a potential 120 million more people, mostly women and children, could be pushed into poverty across the globe by 2030—a crisis that could result in 1 billion climate refugees, displaced from their homes and mostly from black and brown countries.

As many as five and a half million in this nation can buy unleaded gas but can't get unleaded water from their faucets at home. The ecological devastation, from our inner cities to Cancer Alley in Louisiana, reveals the remnants of political pathology, evil economics, and the sick sociology that supported slavery. Four hundred years after the first Africans arrived on these shores, 68 percent of African Americans live within thirty miles of a coal-fired power plant—the distance within which the maximum noxious effects of smokestack plumes are expected to be experienced.

African Americans are 79 percent more likely than whites to live where industrial pollution poses the greatest health danger, and they are overrepresented in populations who live within a three-mile radius of the nation's 1,388 superfund sites. The percentage of African Americans living near the nation's most dangerous chemical plants is 75 percent greater than for the population of the US as a whole. The sins of our past are with us still.

Thus, while the plutocrats practice their centuries-long tradition of racist division, we—poor people, black people, white people, brown people, Indigenous people, Asian American people, Latino people—we, the multiracial majority, have a duty to breathe new life into the other half of the American story, the proud legacy of fusion politics. From Bacon's Rebellion to the abolitionist struggle to the civil rights movement to fusion movements today, there have always been people in this land who rejected the divisions of systemic racism in order to work together for the expansion of democracy.

We must never forget that the abolitionists were free blacks and Quakers, the enslaved and the children of enslavers. The fusion that made a Second Reconstruction possible included students from the North and African American veterans in the South; Jewish rabbis and Christian ministers; labor unions and civil rights organizations; gay organizers and straight sharecroppers. Wedge issues have been used to divide Americans by race, religion, gender, sexual orientation, class, and geography, but fusion politics has always defied those divisions to help us see that we move closer to the promise of a government by the people and for the people when we unite to reconstruct the systems of our common life to serve everybody.

If the extremism of our present moral crises feels overwhelming, it is important to remember that those who are defending the broken systems we have inherited would not be fighting so hard if they did not know that movements for reconstruction and genuine transformation have real power in this moment. By 2040 in the United States, white people will be one among many minority groups. This demographic reality is the driving force behind attacks on immigrants, voter suppression, and the corporate and foreign influence campaigns in American public life. From the US to the UK to Brazil, Russia, China, and Israel, a global economy that was built on the sugar and cotton that enslaved people harvested is threatened by the radical democracy of poor people's movements. Nationalist leaders are responding with a reactionary populism that appeals to racism and xenophobia, deploying the same divide-and-conquer strategies that have been used to resist reconstruction from the beginnings of plantation capitalism.

In this moment, the descendents of those whose hands once picked cotton must join the hands of Latinos, join the hands of progressive whites, join faith hands, join labor hands, join Asian hands, join Native American hands, join poor hands, join wealthy hands, join gay hands and straight hands and trans hands; join Christian hands, Jewish, Muslim, Hindu, and Buddhist hands. When all those hands get together—when the rejected join hands—our togetherness becomes the instrument of redemption. When we join hands in fusion coalitions, we make sure that the promise of life, liberty, and the pursuit of happiness and equal protection under the law and care for the common good will never be taken away or forfeited for anybody, anytime, anywhere.

I choose to believe that the America that has never yet been may nevertheless still be. Together with people from every race, creed, culture, and sexuality, I am committed to building a fusion coalition in the twenty-first century through the Poor People's Campaign: A National Call for Moral Revival, which can be a major impetus toward reviving the heart of this democracy. A commitment to name "ecological devastation" as an interlocking injustice with racism, poverty, the war economy, and a distorted moral narrative is a fundamental commitment of this campaign. We cannot revive the heart of democracy without a Green New Deal, and we can't address the climate crisis without reclaiming democracy and pressing forward together toward a more perfect union.

FOURTEEN

THE NEXT ERA OF AMERICAN POLITICS

GUIDO GIRGENTI AND WALEED SHAHID

"The green dream or whatever they call it." That's what Democratic House speaker Nancy Pelosi had to say in response to Rep. Ocasio-Cortez and Senator Markey introducing the Green New Deal resolution in Congress.

Many pundits were taken aback by the Democratic Speaker of the House undercutting her own colleague's proposal to combat climate change, but it's not surprising that Pelosi thinks the Green New Deal is a fantasy. To borrow the words of Naomi Klein, politicians like Pelosi "have not done the things that are necessary to lower emissions because those things fundamentally conflict with deregulated capitalism, the reigning ideology for the *entire period* we have been struggling to find a way out of this crisis." Pelosi came to Congress in 1987, and like many politicians now in the senior leadership of both parties, her entire career has been built within the confines of the Reagan era, a period defined by neoliberal capitalism and an ascendant Republican Party.

Not only have politicians of this era not taken the actions necessary to lower emissions, but many would-be leaders have been unable to even *speak aloud*, let alone propose and advance, policies that meet the scale of the climate crisis. It's very difficult to imagine a genuinely changed world from inside the Reagan era, even when social and ecological forces are obviously changing *everything* at an accelerating pace.

Neoliberalism has led to many record-breaking achievements: inequality higher than at any time since the Gilded Age; the worst economic crisis since the Great Depression; the longest decline in Americans' life expectancy since World War I; the highest incarceration rate in the industrialized world; the most expensive health care system in the Organization for Economic Cooperation and Development (OECD); the highest levels of student debt; the longest period without an increase in the minimum wage; union membership lower than at any point since the New Deal. And, of course, unprecedented rates of carbon emissions. From the very beginning, racism lay at the heart of this ideology's path to power, and Trump's election confirms that racism is more essential than ever to preserving the power of politicians who serve the rich.

Climate and environmental justice leaders have said again and again that stopping the climate crisis will require confronting the forces of greed and racism that fuel it. Now that neoliberals in both parties have delayed the transition from fossil fuels, and now that the right-wing project of dog whistle racism has culminated in the disaster of the Trump presidency, perhaps more people will listen.

The late Detroit-based organizer and writer Grace Lee Boggs challenged her pupils to look beyond the daily squabbles of politics and ask: "What time is it on the clock of the world?" For politicians like Nancy Pelosi, there is no clock. There is no sense of political eras ending and beginning. There is no reckoning with generation-defining crises. There is no grasping of the historic moment, and its critical importance for the next hundred-plus years of human civilization. There is only an endless negotiation between Team Democrat and Team Republican, an infinite game of bipartisan compromises, with no sense of a larger battle taking place.

Still, the clock is ticking, and change is knocking at the door. Hundreds of years of extractive industry, capped by a ruinous forty-year sprint of escalating profit-seeking, have stirred vast environmental forces beyond our control, and these forces are sending shock waves through human societies around the world. The neoliberal common sense that has dominated our politics for a generation makes less sense every day, and the people of this country are ready for an alternative.

So, what does it mean to end the Reagan era, and forge a new common sense that lays the foundation for the decade of the Green New Deal? To answer that, we need to look back at history.

In the past century, the common sense of American politics changed twice: first, after the New Deal; second, after the Reagan revolution. Historians and political scientists refer to these turning points as realignments. In each realignment, politicians and movements forged a new policy agenda that answered pressing unresolved national crises. The leaders of realignments—let's call them realigners—tie together a majority coalition behind the new agenda, and win power in government to implement that agenda and solidify the new common sense. In building this coalition, realigners link previously unconnected issues and constituencies through a common set of values as they work to establish a new common sense.

We refer to this array of groups and movements that mark a new era as "political alignments," a term coined by veteran organizer and author Jonathan Matthew Smucker. For those of us who work in politics or movement-building organizations, the term "political alignment" is a helpful distinction from the more commonly used "coalition." When we hear about the "New Deal coalition," it's easy to imagine another meeting of progressive groups and Democrats—just at a slightly longer table. But political alignments can't fit around one table. They're so big, according to Smucker, that no single person or group can control their direction; alignments "push the boundaries of what it means to have a cohesive political force" as they work to "mobilize the sympathy and support of a majority."

The project of era-defining realignment is perhaps the biggest goal a movement can aspire to in American politics. It has only happened twice in the past century. Winning a Green New Deal depends on it happening again.

By and large, Americans are told that politics is simply a back-and-forth between Democrats and Republicans, with each party taking a turn responding to the interests of voters. We look to the past and see only alternating periods of red and blue, and this limits our sense of what's possible in the future. From this perspective, there's no hope of sustaining a decade of the Green New Deal; the best we can do is a bit more blue.

It's hard to work toward something that we cannot imagine, and that we don't have the words to describe. Looking back at history through the lens of realignment transforms our understanding of what is possible in the future. The overview of the New Deal era and the Reagan revolution in this chapter is not comprehensive but offers an introduction to the power and longevity of realignments, and how they shape policymaking—of both major parties—for a generation.

This history also offers a few lessons to guide us as we organize toward a new alignment for the next era of American politics. First, realigners must thoroughly discredit the ideas of the old alignment, unshackling politics from the conventions of the preceding era before a new common sense can be established. Realigners must demonstrate how the people in power, their vision, and their policy agenda cannot solve persistent, urgent crises.

Second, realignments are messy. A successful alignment is not a coherent, united alliance; it's broad enough to win national elections, and therefore broad enough to include constituencies with competing and even contradictory interests. Once in power, realigners navigate both the historic opportunity to transform policy and the need to hold together a majority. This tension shapes the priorities and scope of the new governing agenda, surfacing conflicts that can fatally fracture the alignment.

Third, a successful alignment leaves the opposition no choice but to accept the new common sense and accommodate itself to the new policy agenda. In the New Deal era, Republicans governed within the boundaries of the New Deal consensus, even as some of them worked to pull apart the New Deal alignment. In the Reagan era, Democrats have accommodated themselves to the common sense of neoliberalism.

It's tempting to refer to strategy as a "road map," but history does not offer straightforward and legible directions. If winning a Green New Deal requires pulling together a broad alignment of allied constituencies and sometimes competing interests, and yoking that alignment to a transformative agenda for a multiracial democracy that safeguards our freedom, dignity, and prosperity in a climate-changed world, it will require language, ideas, and coalitions that cannot be retrieved, ready-made, from the past. Nevertheless, lessons we glean from twentieth-century realigners provide an essential foundation for today's strategists to build on.

THE NEW DEAL, THE REAGAN ERA, AND THE NEXT REALIGNMENT

The Great Depression of the 1930s was the worst economic crisis the United States and modern capitalism had ever experienced. Almost four full years after the economic crash of 1929, nearly a quarter of Americans remained unemployed. Industrial production had plummeted from $949 million to $74 million in three years. In nearly every major city, Americans with no jobs and no homes set up encampments known as "Hoovervilles," named after the president who presided over the misery. The GOP had dominated American politics for more than a half century, and President Hoover held firm to his party's beliefs that markets self-regulate and government intervention would distort the economy and prolong the crisis.

"Economic depression cannot be cured by legislative action or executive pronouncement," he told Congress at the end of 1930. "The best contribution of government lies in encouragement of . . . voluntary cooperation in the community."

Voters were not convinced. FDR's Democratic Party won by a landslide in 1932, giving Democrats a 196-seat majority in the House of Representatives. A new, broad coalition had come into power, composed of industrial workers, especially Catholic and Jewish immigrants, African Americans, farmers, and the staunchly Democratic South, as well as predominantly middle-class progressive reformers.

Before the New Deal era could begin, the antigovernment agenda of the Gilded Age Republicans had to be publicly discredited. In his first inaugural address, Franklin Delano Roosevelt framed his victory not simply as a rejection of Republicans but as a permanent defeat of the financiers that had caused the crash, and as an end to an economy run by profit-seekers. The Depression persisted, Roosevelt declared, not because of any natural scarcity but "because the rulers of the exchange of mankind's goods have failed," leaving "the unscrupulous money changers . . . indicted in the court of public opinion."

The "New Deal," FDR reflected later, explicitly meant "a new order of things . . . to benefit the great mass" of working Americans and to "replace the old order of special privilege," which now "thoroughly disgusted" the nation. Free to discard old ideas and compelled by immense suffering, the New Deal took as its starting premise government as the guarantor of security and industrial stability. But beyond this premise—a radical departure from the past half century—FDR lacked a singular "transformative vision," instead embracing myriad liberal and progressive philosophies in an attempt to rescue the economy and keep the new alignment together.

The ideas that would define "the New Deal era" did not arrive fully formed with Roosevelt; they were forged in the compromises and conflicts of the next decade. In mid-1933, when FDR sought to stimulate industrial production and boost consumer purchasing through a new National Recovery Administration (NRA), his administration placated every necessary constituency. Corporations demanded protection from trust-busting; labor and congressional progressives demanded the right to collectively bargain; Southern Democrats demanded that agricultural and domestic workers—the vast majority of southern blacks worked as farmworkers or domestics—be excluded from the NRA's jurisdiction, protecting Jim Crow from the threat of enfranchised black workers. Each received a concession, and concessions held the newborn alignment together.

The NRA itself, however, was more than a child of compromise. The keystone of FDR's recovery program, it combined progressive ideas of regulation and labor rights with the tools of economic planning. The agency signaled a profoundly reconstructed economic order: a government-managed industrial economy built on a foundation of business-labor cooperation.

The NRA's balancing act—an overhaul of power in the industrial economy, concessions to coax stable labor-business relations—did not endure. Within a year, socialists helmed two general strikes, shutting down San Francisco and Minneapolis, as an extraordinary wave of labor militancy roiled the country before the 1934 midterm elections. The popularity of Roosevelt alongside this vast mobilized radicalism swept an even bigger Democratic majority into Congress, including freshman progressives who stood to the president's left. FDR's party won, but "the new Congress threatened to push him in a direction far more radical than any" he had considered, William Leuchtenburg writes.

Realigners prepared to seize on the new radicalism in the streets and in Congress. Harry Hopkins, a left-wing New Deal administrator and Roosevelt's confidant, galvanized his staff: "We've got to get everything we want—a works program, social security, wages and hours, everything—now or never." Senator Robert Wagner introduced the National Labor Relations Act in February 1935. Roosevelt proposed a Social Security Act that stalled in the House but soon refocused on reviving a besieged NRA and holding together an alignment under strain.

By 1935, the NRA was failing. Consumers complained of corporations artificially inflating prices; the vague collective bargaining rights emboldened workers to strike and organize but did little to reduce industrial conflict or raise wages; small businesses and big firms sparred over pricing and market share. FDR failed to hold business support. The Supreme Court delivered the death blow, striking down the NRA on May 25, 1935.

Roosevelt had discredited the old order and held together a fraying coalition, bringing even larger majorities into Congress in 1934. But when business opposition and the Supreme Court gutted the heart of the first New Deal, it was the strike waves and the relentlessness of progressive realigners in Congress that made possible the era-defining reforms of the second New Deal: the legal enfranchisement of labor unions, and the first national system of social security.

With draft bills for Social Security and labor rights already in hand, progressives urged Roosevelt to oppose business and pass a new program for labor rights, universal security, and equality. Six weeks after the NRA was invalidated, Congress passed the National Labor Relations Act, and the Social Security Act passed a little over a month later. Southern Democrats remained an essential voting bloc, and again New Dealers compromised with Jim Crow, excluding agricultural and domestic workers from the new programs.

Following these victories, the New Deal appeared open to more radical directions, and realigners in the labor movement believed that the moment was ripe for mass unionization and continued expansion of social and economic rights. John L. Lewis and Sidney Hillman, labor leaders frustrated with the refusal of the American Federation of Labor (AFL) to unionize unskilled workers, formed the Congress of Industrial Organizations (CIO) in late 1935 to "organize the unorganized." As Lewis hired dozens of communists and socialists to build militant unions in the steel, auto, and rubber industries, the CIO allied with Roosevelt to keep the New Deal alignment in power. In mid-1936, they formed Labor's Non-Partisan League and raised an impressive $600,000 for FDR's presidential campaign.

The CIO not only helped reelect the New Deal alignment by funding Roosevelt, it also shifted the balance of power inside the alignment by unionizing African American workers, who soon became essential Democratic Party constituents. The CIO's fusion of racial justice and labor rights into the idea of "New Deal liberalism" sowed seeds that would bloom in the midcentury civil rights revolution.

Hillman and Lewis, both realigners seeking to fulfill the New Deal's transformative potential while maintaining Roosevelt's governing majority, navigated this balancing act quite differently. In his successful crusade to pass the Fair Labor Standards Act of 1938, Hillman proved an effective labor representative in government.

FDR soon appointed him to represent labor in his growing WWII mobilization apparatus. Hillman believed that labor support for Roosevelt's wartime program was essential if labor hoped to shape the postwar economy. When strikes threatened wartime production, Hillman negotiated a no-strike pledge with the AFL and CIO and helped establish a National War Labor Board (NWLB), providing a forum where labor could bargain on equal footing with business and government. The next year, Lewis broke the no-strike pledge, mobilizing miners in protest of the NWLB suppressing wages.

Hillman and Lewis both began as radical realigners, pushing the scope of the New Deal to the left while expanding the ranks of its alignment through a multiracial union. Perhaps post–World War II politics would've been different, historian Alan Brinkley writes, "if leaders more like Lewis than Hillman had managed to shape" unions during the war. Lewis wasn't alone in his growing estrangement from the New Deal. By the mid-1940s, progressive reformers and radical labor leaders had largely lost the fight to expand its agenda. Southern Democrats began to fear that "labor organizing . . . stimulated civil rights activism." So, alongside a business class that FDR believed he needed on his side to win World War II, they limited the legacy of the New Deal to the management of a growing consumer-driven economy while keeping intact the power of private capital.

By the end of the war, the core of the New Deal consensus, the ideas that defined the common sense of American policymaking until the 1970s, had solidified: the federal government protected people from the brutalities of unfettered capitalism by regulating industry, pursued the goal of full employment, guaranteed economic security through social insurance, and invested public money into public goods. Labor unions and collective bargaining were essential to a stable economy, raising wages and mediating conflict with big business. Though there was opposition—some wealthy businessmen began plotting the death of the New Deal before Roosevelt's first term had ended—the growth of the postwar economy affirmed that this consensus worked.

The New Deal alignment established a common sense so dominant that its opponents in the GOP were forced to govern within its limits. In 1954, President Eisenhower scolded conservatives who thought that the policies of the New Deal could be dismantled: "Should any political party attempt to abolish social security, unemployment insurance and eliminate labor laws and farm programs, you would not hear of that party again in our political history." Eisenhower made clear that such a proposal had little heft in mainstream conservatism, since the "tiny splinter group" of anti–New Dealers includes only "a few Texas millionaires and an occasional politician . . . and they are stupid." But right-wing anti–New Dealers were not deterred.

In his 1960 book *The Conscience of a Conservative*, Arizona senator Barry Goldwater denounced mainstream Republicans for openly rejecting the "principle of limited government," singling out an Eisenhower official who conceded that "if a job [must] be done to meet the needs of the people, and no one else can do it, then it is the proper function of the federal government." Four years later, Goldwater defeated the pro–civil rights moderate Republican Nelson Rockefeller in the 1964 GOP primary. Although Goldwater lost to President Johnson in a landslide, the right-wing operatives from Goldwater's campaign became pioneering conservative realigners, searching for ways to oust the New Deal alignment from power.

Meanwhile, progressive reforms made possible by the policies and coalitions of the New Deal continued. President Lyndon Johnson's Great Society expanded the social safety net the New Deal first created, establishing Medicare and Medicaid, federal education funding, and a range of federal antipoverty programs. The seeds planted by the CIO's multiracial organizing in the mid-1930s were growing into a broader Democratic embrace of civil rights. As early as the mid-1940s, "the liberal coalition identified civil rights as a critical front in the battle for economic and social progress . . . [liberals] understood that defeating southern defenders of Jim Crow was essential for liberalism's future," political scientist Eric Schickler writes.

The New Deal era was no golden age. Democrats presided over an expansion of the military and the disaster of the Vietnam War. Discrimination against African Americans in both New Deal and postwar social programs locked in racial inequalities that persist today. Despite this, the movements and policies of the New Deal era moved the nation toward justice and equality. Between 1947 and the early 1970s, the percentage of Americans living in poverty was cut in half. Economic growth benefited the poor more than the rich, as the poorest 20 percent of Americans saw their incomes more than double. The Civil Rights, Voting Rights, and Fair Housing Acts enshrined legal protections against racial discrimination and provided a bulwark against disenfranchisement.

But in the decade following the civil rights victories, the New Deal order unraveled. A conservative counterrevolution seized upon white backlash and economic downturns. Even as President Nixon governed within the limits of New Deal consensus, creating the Environmental Protection Agency and the Occupational Safety and Health Administration, he also pioneered the Southern strategy that conservative realigners would fine-tune for decades to come.

The contradiction at the heart of the New Deal alignment was that it both created the conditions for civil rights victories by empowering a pro–civil rights labor left and depended on compromise with the Jim Crow South to secure governing majorities. Realigners in the labor and civil rights movements like Bayard Rustin actively worked to push Southern Democrats out of the party, while Martin

Luther King tried to organize impoverished Americans of every race in his Poor People's Campaign. But the hope of a labor–civil rights coalition at the helm of the Democratic Party was never fully realized, and a fledgling right-wing alignment was gaining ground. Democrats steadily lost the South without solidifying a new majority themselves.

Civil rights did not simply push reactionary voters into a conservative coalition; right-wing realigners seized on fractures in the New Deal alignment, organizing relentlessly to pull constituencies together into a new alignment. Few realigners saw their task more clearly than the organization-builder Paul Weyrich, or executed it as effectively as he did. Two years before the Reagan revolution of 1980, Weyrich understood his historic analog: "We are . . . very much like the New Dealers of 40 years ago." "With the old coalition dead . . . there is a vacuum," and the task of right-wing realigners, he explained in a 1979 strategy memo, is to "identify the elements of, and form, a new coalition [and] achieve political power through the coalition, which will be guided by the philosophy of individual freedom and personal responsibility."

Weyrich founded the institutions needed to sustain this coalition. When he thought that the Right's think tank, the American Enterprise Institute, failed to counter the liberal Brookings Institution with necessary agility and force, he co-founded the Heritage Foundation. When the IRS stripped all-white church-based schools of their tax-exempt status as a means to enforce racial desegregation, Weyrich saw an opportunity to link the religious Right with "the long-held conservative view that government is too powerful." He told Baptist pastor Jerry Falwell Sr. that "there's a moral majority" waiting to be organized; the same year, Falwell and Weyrich co-founded the Moral Majority, the preeminent organization of the Christian Right.

While realigners like Weyrich seeded key institutions across issues and groups, others dedicated themselves to tying a specific issue and constituency to the ascendant alignment. Weyrich encouraged this. He believed "coalition politics . . . need not mean either compromise politics or party politics," write political scientists Daniel Schlozman and Sam Rosenfeld; groups can "mobilize diffuse voter blocs on the single issues that most mattered to them." These "single issue" realigners, like antifeminist crusader Phyllis Schlafly, were often more accountable to their constituencies and their movement's values than to national leaders like Ronald Reagan, who prioritized maintaining electoral majorities. The tension realigners like Schlafly navigated was between holding the line on a movement's demands while cementing their issue as a key priority within the broader alignment's agenda.

Schlafly became a tribune of social conservatism in the early Reagan alignment, and she balanced supporting Reagan with mobilizing to force her movement's priorities onto his governing agenda. In the fight to stop the Equal Rights Amendment (ERA), Schlafly built the Eagle Forum, a formidable 60,000-plus organization of conservative, pro-family women. Reagan strategists, struck by Schlafly's talent at activating religious and social conservatives, tuned the rhetoric and policies of their campaign to further link abortion, antifeminism, and school prayer to their free market ideology. Schlafly threw her support to Reagan, believing his candidacy was "vindication . . . of the conservative movement" that began with Goldwater. Reagan opposed the ERA and abortion on the campaign trail, but to ensure that these positions didn't cost him the votes of too many moderate women, his campaign announced a Women's Policy Board helmed by two pro-ERA activists less than two months before the 1980 election. Schlafly sent an "emergency telegram" rallying thousands of Eagle Forum activists to call and write the Reagan campaign, prompting Reagan to personally call Schlafly and announce a Family Policy Advisory Board led by social conservatives.

While Weyrich, Schlafly, and the New Right stitched together a winning alignment, the ideas of the New Deal consensus were already losing credibility. Liberals were unable to resolve or explain the economic stagnation of the 1970s with the Keynesian ideas that defined New Deal policymaking, and their consensus was increasingly under attack by a hostile business class. Corporate profits were in decline, and CEOs rushed to find new allies in the Republican Party.

Free marketeers, foreign policy hardliners, white racist reactionaries, and social conservatives came to power through the landslide victory of Ronald Reagan in 1980. *Washington Post* columnist David Broder said that the election "certainly had all the appearances of an era ending—and a new one beginning."

Broder got it right. Reagan, like Roosevelt before him, framed his victory not as a defeat of the opposing party but as a thorough discrediting of the ideas that defined the preceding era. In his first inaugural address, he declared, "In this present crisis, government is not the solution to our problem; government is the problem." That inauguration, historians Steve Fraser and Gary Gerstle write, marked the end of "the New Deal as a dominant order of ideas, public policies, and political alliances."

The economic common sense of the Reagan era—often called Reaganomics, or neoliberalism—is the common sense that has defined politics for nearly half a century, the too familiar straitjacket of "self-regulating markets," privatization, deregulation, tax cuts, union-busting, and pull-yourself-up individualism. Woven throughout Reaganism is the coded language of dog-whistle racism, masking the white supremacy at the heart of the right-wing revolution: real Americans are white Americans; real Americans work for what they get and pay taxes; people of color threaten our country and our economy with violence and laziness. Reaganism wove neoliberalism, strategic racism, and the cultural touchstones of the evangelical Right—anti-feminism, opposition to abortion, and a mistrust of secularism—into the single story of limited government. If government is the problem, they said, then trust the marketplace.

Right-wing realigners had prepared for power, sharpening the policies and ideas they wanted to define the antigovernment Reagan revolution. The Heritage Foundation, Weyrich's brainchild, rushed to deliver a thousand-page neoliberal policy manual, entitled *Mandate for Leadership*, to the president-elect by inauguration. Heritage later boasted that Reagan implemented nearly two-thirds of the manual's policies during his two terms. Aided by neoliberals at Heritage and beyond, Reagan's administration cut taxes, left the minimum wage stuck at $3.35 an hour, and passed broad cuts to public spending. Gunning for the heart of the New Deal alignment, Reagan fired over 11,000 striking air traffic controllers, signaling to employers that the government would do little to enforce existing labor protections. The number of strikes plummeted; reported violations of the Wagner Act skyrocketed. Between 1979 and 1988, the number of Americans living below the poverty line rose from 26 million to 31.5 million. Meanwhile, the incomes of the top 1 percent grew each year, and by the end of the decade, they held 39 percent of the nation's wealth. In coalition with Wall Street, Reaganites removed New Deal protections that safeguarded families from taking on too much household debt, while removing limits on risky investments big banks could make—setting us on the path that led to the crash of the housing market and the 2008 financial crisis.

Despite all the cuts and deregulation, the "small government" agenda never really meant a weak, unintrusive government; the administration would rewrite the rules of the economy to benefit big corporations and the wealthy, while expanding the US military, prisons, and law enforcement programs targeting people of color. Reagan's war on drugs, announced in 1982, sparked an explosion in law enforcement spending. Between 1981 and 1991, the Department of Defense's anti-drug program budget increased from $33 million to $1,042 million; the Drug Enforcement Agency's anti-drug allocations grew from $86 to $1,026 million. In selling these programs to the public, the dog whistling never stopped. Reagan's team worked hard to publicize "welfare queens" and the crack cocaine epidemic of the mid-1980s, ensuring the media coverage was blanketed with images of black crack addicts. During Reagan's presidency, the US prison population doubled from 315,974 to 739,980. It continued to increase, plateauing at around 1.5 million between 2010 and 2020.

Under Reagan, neoliberalism and "tough-on-crime" politics did not placate everyone in the new alignment, and social conservatives continued to fight to ensure that their issues remained a priority in the administration. Again seeking to hold moderate women in his coalition despite his embrace of the anti-ERA movement, Reagan launched a Fifty States Project to identify and repeal "any remaining discriminatory laws" against women "at the state level." Schlafly opposed this project, pushing the administration to refuse any concession to women's rights. Reagan did his best to signal that neoliberals and social conservatives had an equal place in the governing alignment: "We do not have a separate social agenda. We have one agenda. Just as surely as we seek to put our financial house in order . . . so too we seek to protect the unborn." But his nomination of moderate Sandra Day O'Connor to the Supreme Court, a fulfillment of a campaign promise to nominate a woman, infuriated pro-lifers who'd believed that Reagan-appointed justices would work to strike down *Roe v. Wade.* "We are a movement in disarray," one pro-life movement told the *New York Times* in 1981. Weyrich, always a keen observer of coalition dynamics, noted the competing interests in Reagan's alignment: "the social issues aren't big in the Country Clubs."

Right-wing realigners, like New Dealers before them, navigated the tensions of coalition-building and transformative changes in policy. They successfully "reshaped the national agenda," as the *New York Times* said at the end of Reagan's first term. Shrinking the government, freeing the market, cracking down on crime, defending traditional family values—these were the goals that defined sensible politics. The center of gravity that both parties orbited had moved far to the right, and Reagan knew it. In his farewell address, he said that while pundits referred to his administration's accomplishments as the Reagan revolution, for him "it always seemed more like the Great Rediscovery: a rediscovery of our values and our common sense."

This common sense endured: "Each of Reagan's successors has been subject to political expectations that he set," wrote political scientist Stephen Skowronek two decades after Reagan's victory. "Politically speaking, America is still working its way through the Reagan era."

After Jimmy Carter lost all but four states to Reagan in 1980, Democrats needed to decide what path they would take in Reagan's America. A faction of Democrats began working on a strategy that would break with the principles and allegiances of the New Deal era. Some of the intellectual leaders of this movement embraced the moniker of neoliberalism. In a widely read 1982 op-ed entitled "A Neo-Liberal's Manifesto," magazine editor Charles Peters stated that "we no longer automatically favor unions and big government or oppose the military and big business." These ideas found a home in the Democratic Leadership Conference, chaired by Bill Clinton in the late 1980s. Progressives balked, and, in campaigns like Jesse Jackson's "Rainbow Coalition," sought to win power with the multiracial fusion politics that had animated labor and civil rights in the New Deal alignment. Progressives like Jackson lost the fight inside the Democratic Party, and, soon enough, the neoliberals were calling themselves New Democrats, seeking a "third way" between progressive and conservative policymaking.

With Bill Clinton as their standard-bearer, the New Democrats took the White House in 1992—and like Eisenhower after the New Deal, they accepted the common sense of the new era. This Third Way would adopt the Right's "tough on crime" dog whistles. Before the New Hampshire primary in 1992, Governor Bill Clinton oversaw the execution of Ricky Ray Rector, a mentally impaired black man who, legal scholar Michelle Alexander writes, "had so little conception of what was about to happen to him that he asked for the dessert from his last meal to be saved for him for later." Following the execution, Clinton said, "No one can say I'm soft on crime." Two years later, President Clinton signed the Violent Crime Control and Law Enforcement Act, mandating life sentences for three-time offenders. Over his two terms, the US prison population rose nearly 60 percent.

Clinton's Third Way also continued Reagan's work of repealing New Deal consumer and financial protections and rolling back social programs. In his 1996 State of the Union address, President Clinton declared that "the era of big government is over." Seven months later he dismantled the Aid to Families with Dependent Children (AFDC) program created by FDR's Social Security Act and replaced it with limited welfare benefits tied to strict work requirements. Three years after his pledge to "end welfare as we know it," Clinton approved the repeal of the Glass-Steagall firewall between commercial and investment banking that FDR had passed in 1933, leading to the rise of even bigger banks and making the 2008 crisis "broader, deeper and more dangerous."

In the following decade, Reaganism, now carried forward by the tax-cutting, warmongering George W. Bush, was threatened by its own failures. The criminally negligent and racist government response to Hurricane Katrina, the disastrous Iraq War, his failed attempt to privatize Social Security, and the 2008 financial crisis made one thing clear: the people in charge and their commitment to "small government," imperial military intervention, and "free markets" were driving the country to ruin.

Elected in the wake of these disasters, President Barack Obama elicited talk of generation-defining change—a "New Deal": "Everywhere these days, the comparison is being made: Barack Obama in 2009 and Franklin Delano Roosevelt in 1933," said journalist Adam Cohen. But Obama did not intend to undo the Reagan consensus and replace it with something new; his approach hewed much closer to Clinton. Obama praised Clinton's Third Way for trying to transcend ideological conflict and for recognizing "how markets and fiscal discipline could help promote social justice . . . that not only societal responsibility but personal responsibility was needed to combat poverty." He told *60 Minutes* soon after his victory that he hoped not to "get bottled up in a lot of ideology and 'is this conservative or liberal?,'" saying that whether an idea came "from FDR . . . or from Ronald Reagan, if [it's] right for the times then we're going to apply it."

Coming into office amid the worst economic crisis since the Great Depression, with Wall Street on its knees, Obama had an opportunity to remake the American economy, with Democrats controlling both the House and the Senate. But his team, led by Clinton administration veteran and economist Larry Summers, intended to restore the economic and financial system as it had been pre-crisis, and not stray too far outside the norms of Reagan-Clinton economic policy. The goals of the stimulus, as an internal administration memo reveals, were to remain committed "to a responsible budget," stabilize the banks, and end the recession without the risk of "an excessive recovery package . . . spook[ing] markets."

When Christy Romer, an economist on the transition team, proposed a $1.7 trillion stimulus—big enough to bring employment back up to pre-crisis levels—Summers made clear he would not present such a high figure to Obama. When Romer cut her proposal down to $1.2 trillion—in her mind a sensible middle ground between what Obama's team believed was politically realistic and what the economy needed to recover—Summers again removed the proposal from the memo Obama ultimately received. All trillion-dollar proposals were not practical, in his mind. When a member of the transition team proposed chartering new green banks to finance clean energy development and energy efficiency projects, Summers demurred: "the problem is that you are talking about creating more debt." When Obama—searching for a big moonshot to boost the public reception of the stimulus—proposed overhauling the nation's power grid to expand renewable energy, Summers spurned meddling in the market structure. "The government's job is to remove regulatory obstacles," not fund and direct overhauls, Summers advised. When Obama kicked off the transition team's economic work, Treasury Secretary Timothy Geithner set guidelines for his policymaking: he would not want to punish bankers, or even put their bonuses in jeopardy.

With the moonshots set aside, the 2009 stimulus, totaling $831 billion, expanded antipoverty programs, invested $90 billion in clean energy—a genuinely massive boost to US solar and wind at the time—and gave out $288 billion in tax cuts to individuals and businesses. Over 30 million Americans would lose their jobs in the course of the Great Recession; 9.3 million Americans lost their homes.

While rescuing the economy from depression, the Obama administration, like Clinton's before him, largely accepted the boundaries of Reaganism, especially in his first term, even as it occasionally reflected the spirit of a new progressive era. The stimulus itself is a metaphor for an administration caught between two eras. Forced to manage a crisis caused by deregulation and corporate fraud, the stimulus expanded welfare programs and invested in green jobs, yet refused to stray too far from the Reagan consensus: don't spook the markets, don't increase the deficit too much, stabilize the banks but don't restructure them, and cut taxes. Likewise, when Obama's Democratic Party pursued health-care reform, they expanded Medicaid, a signature program of the Great Society, while also creating more marketplaces for insurance companies to sell to individuals, a policy idea borrowed from the right-wing Heritage Foundation.

Liberals can argue until the end of time about whether the Affordable Care Act and the stimulus were as progressive as they could've been, given Republican obstruction in Congress and the Clintonite centrists in Obama's own party. However, we won't know, because Obama's approach was not to fight against the Reaganites and Clintonites for progressive, transformational change. His goal was not to end the Reagan era but to transcend partisanship. Regardless of his intent, there was no big social movement or caucus of rowdy progressive legislators to back him up. To use Sunrise's language, progressives to the left of Obama had little people power or political power in 2009. Meanwhile, the right-wing still had Fox News, armies of corporate lobbyists, a network of conservative think tanks, and soon a Tea Party movement, all arrayed against the president.

Following health-care reform and the successful stonewalling of climate legislation by right-wing and corporate opposition, Obama turned to a central goal of the Reagan revolution: deficit reduction and "entitlement reform." In 2010, he created a National Commission on Fiscal Responsibility and Reform, which recommended cutting social programs and raising the retirement age. During a grand bargain over the budget in July 2011, he offered Republicans cuts to Social Security, Medicare, and Medicaid in exchange for raising taxes on the wealthy. In August, less than three years after Wall Street crashed the economy, Obama called for "shared sacrifice" to reduce the debt. Despite his efforts, the national mood would not remain focused on balanced budgets and bipartisan compromise.

That same summer, over a thousand activists got arrested outside the White House, demanding that Obama reject the Keystone XL oil pipeline. A month later, Occupy Wall Street changed the national conversation from "cut the debt" to "tax the 1 percent." The following year, mass marches sprang up to protest the murder of Trayvon Martin, demanding justice for black lives. Undocumented Dreamers sat in at Obama campaign offices until the president announced protections for undocumented Americans who immigrated as children. Workers went on strike demanding a $15 minimum wage. The Black Lives Matter movement grew as the epidemic of police killing unarmed black Americans continued unabated.

In the protest movements of the Obama era, we saw the seeds of a new progressive force in American politics. By mid-2013, journalist Peter Beinart was already noting the "rise of a new New Left"—a millennial generation shaped by the financial crisis, less supportive of capitalism than any other generation, and set to become the biggest voting bloc in America. "The door is closing on the Reagan-Clinton era," Beinart predicted.

Three years later, a seventysomething democratic socialist won 43 percent of the Democratic vote in the 2016 Democratic presidential primary. Meanwhile, a racist former television personality won the Republican nomination—and his path to that nomination revealed a potential fracture in the Reagan coalition. The free marketeers want unjust free trade agreements and cuts to entitlements, but Trump railed against NAFTA and vowed to protect Medicare and Social Security; the foreign policy hawks want to preserve America's right to intervene abroad, but Trump, unlike his fellow Republican candidates, criticized the Iraq War vehemently. While rejecting Reaganite economic and foreign policies, he turned the dog whistles of the Reagan era into a bullhorn of xenophobic, racist nationalism.

Political scientists wondered if his nomination meant that Trump might "realign" the parties, as the Reagan alignment's fusion of big business and racist populism split apart. "Trump's presidency signals the end of the Reagan era," wrote political scientist Julia Azari after the 2016 election. In office, however, Trump has not bucked the Republican establishment or forged a new common sense to replace Reaganism. Despite that, Reaganism appears weaker than ever.

Since Reagan's landslide victory in 1980, every subsequent GOP president has won a smaller share of the popular vote, concluding with Trump at 46 percent—losing the popular vote to Hillary Clinton. Trump, and the GOP as a whole, could only clinch this second-place finish by refusing to make neoliberalism the center of his campaign. With control of the House, Senate, and White House in 2017, the GOP had no broad popular agenda to advance, just one more deeply unpopular tax cut and one more unpopular failed attempt to repeal Obamacare. Their vision is exhausted, in large part because they've implemented much of it these past four decades. With the labor movement in decline, curtailed abortion rights, the deregulation of multiple industries, economic inequality reminiscent of the Gilded Age, and racial resegregation, the Reaganites have little left to offer.

Meanwhile, the Democratic Party is in the midst of a battle for its soul. Bernie's "radical" ideas—Medicare for All, free college, a

Green New Deal–sized climate plan—are no longer just Bernie's but define the agenda of a rising progressive faction in Congress and are increasingly the center of gravity in national Democratic politics. Every 2020 Democratic presidential candidate was expected to respond to Bernie's ideas; no one was asked if they supported Bill Clinton's Third Way. Two out of three leading candidates for the 2020 Democratic presidential nomination called for a break from the Reagan era, either through "big, structural change" or "political revolution." It's also revealing that the front-runner, Joe Biden—whose political career epitomized Clinton's Third Way—is not defining his candidacy with an embrace of free markets, fiscal discipline, and government downsizing. Rather, Biden's platform has moved far leftward, embracing policies like a robust public option and a $17 trillion clean energy investment.

The retreat from the Reagan consensus on the national stage reflects a leftward-moving electorate. "Public support for big government—more regulation, higher taxes, and more social services—has reached the highest level on record," reported journalist Matt Yglesias, after a new study that measures public opinion on core economic questions—size of government, taxes, regulation—found the most liberal national mood since 1961. Millennials, in particular, remain a starkly progressive generation and will be an indispensable bloc of voters in years to come. In 1984, Reagan handily won the youth vote; today, young Americans like us are voting for socialists and progressives, and we love the Green New Deal.

The Democratic Party establishment is still powerful, cautioning "moderation" and dictating much of what happens in the House of Representatives. However, some neoliberals have begun to concede that a new era and a new politics is necessary. Brad DeLong, a Clinton administration economist who worked on the neoliberal reforms of the 1990s, lamented the failure of the Third Way's accommodation with Reaganism: "Over the past 25 years, we failed to attract Republican coalition partners, we failed to energize our own base, and we failed to produce enough large-scale obvious policy wins to cement the center into a durable governing coalition." Though the media and dishonest Republicans deserve some blame, "shared responsibility is not diminished responsibility," DeLong wrote. "And so the baton rightly passes to our colleagues on our left. We are still here, but it is not our time to lead." Ed Kilgore, a former vice president at the Democratic Leadership Council—the birthplace of Clintonism's Third Way in the late 1980s—echoed DeLong in a piece headlined "A New Role for Democratic Centrists: Helping the Left Win." Journalist George Packer underscored the point in *The Atlantic*, wondering whether Americans, no longer satisfied with "fine-tuning a grossly unjust economy and a corrupt political system," are "ready to hear radical solutions." If so, he wrote, "perhaps this means a realignment of the party and the country to the left."

While these shifts in American politics began to quicken, the world's leading scientists were sounding the alarm: "rapid and far-reaching transitions . . . unprecedented in terms of scale" across major sectors will be required to keep the world below 2 degrees of warming. Another report, another call for massive transformation: "Global greenhouse gas emissions must begin falling by 7.6 percent each year beginning 2020—a rate currently nowhere in sight" if we want a chance at staying below 1.5 degrees of warming. A few months before this report was released, a young democratic socialist from Queens ousted a Wall Street–friendly Democrat and soon, alongside a burgeoning youth movement, brought the Green New Deal onto the national agenda.

What happens next?

BECOMING REALIGNERS

What happens next depends on who forges a new alignment that can win power with a vision that responds to the crises of our time.

Reaganism seems unable to revive itself. As political scientist and constitutional law scholar Jack Balkin wrote in 2019, "The [Reagan] regime is now dying, but a new regime has yet to be born. What the new regime will look like is still unclear." The climate crisis demands massive government-led transformations, but that does not mean that a progressive realignment is inevitable. In Europe, far-right parties acknowledge the climate crisis and use it to argue that EU nations should focus only on protecting people of European descent and keep out everyone else. "Borders are the environment's greatest ally," a young member of France's far-right National Rally declared in April 2019. This is the terrifying alternative to progressive realignment, a competent Trumpism that reformulates plutocracy, white supremacy, and imperialism for a world of heat waves and rising tides. Balkin notes that a right-wing alignment mustering a "second wind" is not unprecedented: Gilded Age Republicans did exactly that when they defeated the populist progressive insurgency of William Jennings Bryan and sustained their dominance for another three decades, until Hoover's defeat in 1932.

Winning a Green New Deal requires replacing the dying Reagan regime with a progressive alignment. Realignments are messy, rare, and big. No single person or group has their hand entirely on the rudder, and there's no step-by-step method to succeeding.

Lacking a road map for realignment, we can take a few lessons from history. First, we cannot avoid the battle of ideas. Both major realignments of the past century began with a direct and public rejection of ideas that dominated the preceding era. Progressive realigners in Congress and mobilized workers, through draft legislation and mass strikes, brought the ideas of labor rights and universal social insurance to President Roosevelt when his administration was floundering after the death of the National Recovery Administration. Paul Weyrich seeded a vast right-wing coalition without accepting the terms of the New Deal, instead building institutions like the Heritage Foundation to define the next era.

Second, there is no avoiding the question of race and racism in American politics. The New Deal alignment was undone by its contradictions on the issue of civil rights. Right-wing realigners understood that race would remain central to American politics—and they wielded racism as a tool of mobilization, coalition-building, and coalition-breaking. Winning a Green New Deal requires a political common sense and an alignment that cannot be torn apart by the Right's strategic racism. Defeating the Right's racism requires a renewal of social solidarity across all forms of difference. Solidarity is the bond that holds us together and gives us a sense of shared fate, in particular against the forces of racial division. If a new, progressive alignment contains some leaders who question the importance of racial justice, we should follow the example of the realigners in the CIO and the leaders of the civil rights movement: organize a bigger multiracial movement, build institutions of multiracial working-class power, and change the balance of power inside the alignment.

Lastly, we will need to navigate the tension previous realigners faced: seizing the opportunity for transformative policy change while keeping the alignment big enough to win governing power; adhering to our highest principles while persuading new allies. But, to borrow Weyrich's insight, we should not assume that coalition politics automatically leads to compromise. Weyrich broadened the alignment by weaving the demands of broad movements and recently activated constituencies, like the newly politicized evangelicals in the Moral Majority, into a coherent right-wing agenda centered on limited government. However, both Reagan and Roosevelt disappointed the leaders of grassroots movements, compromising on certain promises or stalling specific policies out of fear that their electoral coalition would fall apart. Movements are always needed to widen the possibilities of the political agenda and make certain compromises unacceptable.

Different organizations and leaders will play different roles. Like Phyllis Schlafly and the Eagle Forum, or John L. Lewis and the CIO, some realigners will focus relentlessly on organizing the biggest grassroots constituencies possible. They will mobilize those constituencies both to elect allied politicians and to push the policy agenda toward their demands, sometimes threatening national politicians like Roosevelt and Reagan with strikes, boycotts, and bad press. Some may choose the path of Sidney Hillman in the early 1940s, resolving conflicts between groups with the hope of protecting a movement's place inside the governing alignment. We may find ourselves cursing the choices our fellow realigners make in navigating these quandaries, as John L. Lewis cursed Sidney Hillman. We should hope to have these kinds of problems, because these are the problems of an alignment that has won power.

FIFTEEN

FROM PROTEST TO PRIMARIES

The Movement in the Democratic Party

ALEXANDRA ROJAS AND WALEED SHAHID

BUILDING JUSTICE DEMOCRATS: ALEXANDRA

My grandparents, who immigrated to the US from Peru and Colombia, raised me in East Hartford, Connecticut. One of my earliest memories is traveling to visit my family in Lima. I remember walking up to the family home, a thin concrete building that wasn't fully finished. The bedroom where my cousin slept was covered by a tarp, and I couldn't help but feel guilty. *What makes me so different from my cousins?* I wondered. I knew my grandparents were working hard to provide a security that they never experienced growing up.

After high school, I moved to California and took three jobs so I could pay for community college—two jobs selling clothes that I couldn't afford at fancy retail stores (a pair of tights cost $89!) and a third organizing legal documents for an attorney. My goal was to maintain a 4.0 GPA, qualify for in-state tuition at UC Berkeley or UCLA after working in California for a few years, and make my family proud.

In fall 2015, a friend sent me the Bernie Sanders campaign

announcement video. Standing on the shore of Lake Champlain in Vermont, Bernie said things I had never heard a politician say in this country. I'd seen my friends go into crippling student debt and still struggle to provide for their families, despite their degrees. I'd felt powerless when my own family members were struck by illness and couldn't afford treatment. I felt like this was more than a campaign for president—it was about resetting the direction of our politics to change the lives of working people like my family.

Over the next few weeks, I started volunteering as much as I could for Bernie. I set up tables at other colleges in Orange County to get the word out. I learned what the word "barnstorm" means.

Then one day I found out I wasn't eligible for in-state tuition at UCLA or Berkeley. Being a dependent of a parent in another state disqualified me—and now I was on the hook to pay twice as much as I'd budgeted in out-of-state tuition.

So I left. In January 2016, I quit my jobs, put university on hold, and moved to Burlington, Vermont, to work full-time for the Bernie campaign.

Luckily, I was far from alone in dropping everything to fight alongside Bernie. Leaders emerged by the thousands from bars and restaurants, from community college classrooms, from factories and big-box stores across America. I was stunned by how many people were ready to organize for an old democratic socialist.

Bernie's historic grassroots campaign, nearly winning the Democratic primary against all expectations, showed what is possible when working people stand up and fight. Talking to supporters from around the country, I realized that people were deeply concerned about what came next. The question was always: "No matter who becomes president, how are they going to get anything done in Congress?" Not only would they be legislating with Republicans and playing defense against Fox News, the establishment wing of the

Democratic Party would not push for a progressive agenda. A group of us from the Sanders campaign saw an opportunity to unite people around a movement like Bernie's, focused on electing a brand-new Congress to win a progressive agenda.

If we could build a diverse slate of candidates that truly represented all parts of our society and economy from communities across America, we could transform the Democratic Party.

This is why we created Justice Democrats (JD). The only way we can build a Democratic Party that works for all people, not just wealthy corporate donors, is by investing in women, people of color, and working-class leaders who actually face the problems they would be trying to solve in Congress. These candidates represent the voters who are increasingly becoming the base of the Democratic Party. Their politics and commitments to a progressive agenda come from their communities and movements, which give life to their campaigns.

Initially we thought that quantity was the key to inspiring people. We set our aims high, trying to raise millions of dollars to challenge hundreds of incumbents. Everyone said we were crazy—"You're not going to elect dozens of people to Congress!" Turns out they were right.

Not only is it hard to convince working-class people, especially women of color, that they can run for Congress, but it's incredibly difficult to recruit, support, and manage multiple congressional campaigns simultaneously.

Then Alexandria Ocasio-Cortez, a twenty-seven-year-old Latina bartender from a working-class family in the Bronx, was nominated by her brother through the Justice Democrats nomination form in 2016. I still remember reading his letter. Our team could see how much she cared for her family, how dedicated she was to building an America for all of us.

Alexandria had incredible grassroots energy, both online and in her district. She was running against Joe Crowley, a twenty-year incumbent and the fourth most powerful Democrat in the House. Not only was Crowley bought out by big corporations and real estate developers that hurt working families like Alexandria's in the Bronx and Queens, he didn't reflect the diversity of his district or the urgency of the moment. Alexandria was the ideal candidate, someone who could both mobilize the progressive base of the Democratic Party and reach voters not expected to turn out for a typical midterm primary.

Similar to Bernie's campaign, this primary felt like a referendum on the soul of the Democratic Party and, in turn, the soul of America. We had recruited a millennial, working-class challenger to run against a career politician whose apathy and corporate-backed politics directly affected her family and her community. Alexandria knew what was at stake for her district, and she had nothing to lose.

Her win set off an earthquake that shook our party and our country. It was unimaginable until it happened. With no ties to the Democratic Party, she took on and defeated one of the most powerful senior Democratic leaders in the country.

Three more Justice Democrats were elected that cycle. Ayanna Pressley defeated another white male incumbent in a diverse district, and Ilhan Omar and Rashida Tlaib became the first two Muslim women ever to be elected to Congress. The #Squad was headed to Washington, DC. They weren't supposed to win. But the Democratic Party is changing, from the bottom up.

There is a new, diverse, working-class generation of leadership that is coming to Congress. As Alexandria has said herself, for one of us to make it, one hundred of us have to try. It will take a multiracial, multigenerational mass movement of working people in districts across America to get it done. But now we see what is possible. We won, and we will do it again.

THE DEMOCRATIC PARTY WE NEED: WALEED

We're a generation in a hurry—looking to make change, and make change fast. In the Obama era—from Occupy to the fight against the Keystone XL pipeline and the rise of movements like the Dreamers and Black Lives Matter—it felt like our movements were winning battles here and there but losing the war. We were winning over public opinion on a range of issues but were failing to translate public opinion into legislation and lasting political power.

On issue after issue, we found ourselves fighting the same entrenched Democratic Party establishment that took money from Wall Street, wanted to build more pipelines, promoted trade deals that labor unions opposed, looked the other way when it came to record levels of deportations, and were afraid to take on corrupt police departments.

If Democrats were so often the party of the status quo, then where was our party?

I came to realize the importance of an inside-outside strategy—the blending of protest and electoral politics—after personally witnessing the power of such an approach at my first job after college. After a dozen or so job rejections during my final semester, I joined AmeriCorps and worked at a legal and social services nonprofit serving low-income immigrants.

Twice a week, our organization held "open office hours" in the evenings so that immigrants could receive a free consultation regarding their legal issues. I would run through a questionnaire with them, asking: "When did you arrive in the United States? How did you get here? Where do you work? Do you have any family here? What languages do you speak?"

I've been here for ten years. I overstayed my visa because I started a life here. I work on a farm—chickens, mushrooms. My only family is my wife and kids. I speak Kiche, an indigenous language, and Spanish.

By the end of checking off boxes on my clipboard, I would eventually have to deliver the line I had been trained to give to about 80 percent of the immigrants who walked through our doors. "I'm sorry. There's not much we can do here. There's no pathway for you to receive relief or documentation under the current law. Please watch CNN. President Obama may pass a new law this year."

Increasingly polarized political parties working in a gridlocked Congress would decide whether there would be a pathway to legal status for 11 million undocumented immigrants. In June 2013, an immigration reform bill finally passed in the Senate, with fourteen Republicans joining Democrats. House Speaker John Boehner, facing pressure from conservatives aligned with the Tea Party movement, refused to bring the Senate bill to a vote.

"Too many members worried that potential Tea Party primary opponents, or Democrats in November, could use their votes on immigration against them," said *Time* magazine.

Despite all the gains that the movement of Dreamers—undocumented youth brought to the United States by their parents—had made since 2008, comprehensive immigration reform was hanging by a thread.

Near the end of my time at the nonprofit, the debate on immigration reform came to a swift end. Tea Party activists were backing a primary challenger to House Majority Leader Eric Cantor. Fox News anchor and talk show radio host Laura Ingraham tweeted: "#VA07 Congressional race is ground zero to stop #amnesty. Dave Brat fights to protect your job and wages." On June 10, 2014, Dave

Brat defeated Eric Cantor in a stunning upset in which the challenger was outspent nearly 25 to 1.

"Cantor Loss Kills Immigration Reform," read the *Politico* headline the next morning.

Cantor spent $5 million and had twenty-three paid staff members. Brat spent less than $200,000 and his campaign manager was just twenty-three years old. One small race had changed the entire political calculus on immigration in the Republican Party and in the nation. None of the families I had told to keep watching CNN would probably ever know about Dave Brat. But they would feel the effects.

The conservative movement had successfully created an inside-outside strategy to achieve its goal of killing immigration reform and opposing President Obama's agenda. A combination of targeted protest and a handful of successful primary challengers made it increasingly painful for Republican politicians to break with their party's base.

A year later, the Democratic Party was experiencing its own earthquake. A wiry-haired senator from Vermont who called himself a democratic socialist launched his campaign. I was one of the thousands of people in my generation who was inspired by Sanders's call for a "political revolution" and joined his campaign.

During our door-knocking trainings, I would ask our volunteers why they were supporting such a long-shot campaign.

- *I'm for Bernie because I support the Fight for $15.*
- *I'm against the Keystone XL pipeline.*
- *I support Black Lives Matter and am appalled by Hillary Clinton's "superpredator" comments.*
- *I think President Obama deported too many immigrants, and I'm with the Dreamers.*
- *I'm with Bernie because he doesn't take Wall Street money.*

Something massive was happening. These volunteers had been inspired by the social movements that had grown during the Obama era and were now bringing that energy into an electoral campaign.

But, movement-sparked inspiration aside, virtually none of the volunteers I worked with on the Sanders campaign belonged to any social movement organization. Instead, they were part of the thousands of ordinary people who get involved in electoral politics whether through voting or volunteering for a campaign. My parents, for instance, would probably never participate in a confrontational sit-in, but they were part of the wave of new immigrant and Muslim voters who helped propel Barack Obama to victory over Hillary Clinton in the 2008 presidential primaries.

If we could bring the coalition that Sanders helped ignite in 2016 into Democratic primaries in down-ballot races and expand it, we could create a Democratic Party that reflected its voters, not big corporate donors.

This was looking like what the Tea Party had successfully done within the Republican Party: bring the movement's outside protests and ideas inside the party by running movement-aligned candidates in primaries. For progressives, the biggest lesson to be learned from the Tea Party's playbook is that they didn't work for the Republicans—they made the Republicans work for them.

While the Tea Party had brought out the worst aspects of the Republican Party's racism and corporate greed, the Sanders campaign was attempting to restore the Democratic Party to some of its redistributive roots in the New Deal and the Great Society. For decades, a "neoliberal consensus" had taken hold in the Democratic Party, most famously embodied in President Bill Clinton's claim that "the era of big government is over." Sanders's campaign represented a renewed faith in the ability of government—not big corporations—to solve problems in America. As Alexandria

Ocasio-Cortez later put it, "When people try to accuse us of going too far left—we're not pushing the party left. We are bringing the party home."

On the national stage, running for president, Sanders revealed a constituency of voters that Occupy never could. There are only so many people who will show up for a sit-in. There are many more people who will show up to vote, let alone knock on doors or phone-bank, for a candidate aligned with the issues that sit-ins might call attention to. And by the end of the campaign, nearly 43 percent of voters who participated in the Democratic primaries supported the democratic socialist calling for a revival of President Franklin Delano Roosevelt's "Second Bill of Rights," including the rights to health care, housing, education, a living wage, and more.

But his campaign did not represent the ceiling of what was possible in transforming the Democratic Party into a vehicle for lasting social change. Sanders, a septuagenarian from one of the whitest states in the country, won more votes from millennials than Trump and Clinton combined, including a majority of women and people of color under forty-five years old. If Sanders could do that without having changed his message over the past four decades, how popular could this progressive vision be if it was taken up by young leaders who represented the growing base of the Democratic Party: people of color, women, and downwardly mobile millennials?

As the campaign continued through the spring of 2016, Sanders struggled to earn the support of key constituencies in the Democratic Party, namely older African American voters. In his 2020 campaign, Sanders made enormous strides in tackling these vulnerabilities, but as the Democratic Party was becoming younger and more diverse, organizers began to see an opportunity for electing a new generation of leaders who represented the growing progressive portion of the party's base both ideologically and demographically.

WHO'S MOVING THE PARTY?

The death of immigration reform at the hands of the Tea Party rebellion revealed, as journalist Mehdi Hasan likes to say, Democrats bringing a knife to a gunfight while Republicans bring a rocket launcher. In fact, over the past two decades, conservatives have conducted far more primary challenges than the progressive Left. This is one cause of what scholars have called *asymmetric polarization*: why Republicans in Congress have moved much more dramatically rightward than Democrats have moved left.

"Despite the widespread belief that both parties have moved to the extremes, the movement of the Republican Party to the right accounts for most of the divergence between the two parties," wrote Princeton political scientist Nolan McCarty.

Tea Party–aligned candidates failed in the vast majority of their primary challenges to incumbent Republicans. But the perception scared moderate Republicans to death. And when a challenger could eat into the incumbent's share of the vote—say, from 100 percent to 60 percent—they would reveal that 40 percent of the primary electorate wanted more conservative representation. Today many Republicans are more afraid of a primary challenge than a general election, while most Democrats, before Ocasio-Cortez's victory, haven't been afraid of primaries at all.

The Tea Party movement pressured and provided political cover for Republican leaders to make more unreasonable demands and use more aggressive tactics in opposing President Obama. But the political dynamic led to Democratic leaders sometimes bending over backward to assuage the concerns of the far-right insurgency. Feeling the intensity of the Tea Party inside-outside strategy, President Obama even put forward Republican-style proposals to cut Social Security and Medicare by $200, to $380 billion in 2013.

Progressive social movements capturing the hearts and minds of the nation needed to create a counterweight to the Tea Party and needed to move the Democratic Party toward a commonsense agenda to fix our country's major problems. But how should young people and progressives seeking to influence the Democratic Party get their voices heard in a system corrupted by big money and establishment gatekeepers?

WHAT IS THE DEMOCRATIC PARTY, REALLY?

You might think that the best way to influence the Democratic Party would be to join the College Democrats on your campus or become involved in your local party. But power in the Democratic Party doesn't fully reside within any "official" Democratic Party organization.

In fact, we need to first challenge the assumption that the Democratic Party is a singular, coherent thing. It's not a team, it's the arena.

Who controls and sets the direction of the Democratic Party? Is it Barack Obama? Is it House Speaker Nancy Pelosi and Senate Minority Leader Chuck Schumer? Is it Democratic National Committee chairman Tom Perez? Is it major donors like Tom Steyer and George Soros? Is it major institutions like labor unions and Planned Parenthood? High-profile consultants like David Axelrod and James Carville? Think tanks like the Center for American Progress?

The truth is that there is no singular organization or individual that controls the Democratic Party. Politicians, organizations, and donors all jostle with one another to exert influence over the direction of the party on a number of any given issues. This is often referred to by political scientists as a *weak party system*, meaning that the official party apparatus has a limited amount of control over the direction of the party. Political parties are the field on which the game is being played. The players, on the other hand, are the organizations, movements, donors, and politicians all competing to be in the starting lineup.

WHY PRIMARIES MATTER

Party politics in the United States is quite unlike party politics in many other countries. Our parties don't really have "members" who come together and decide the direction that everyone who is a Democrat or a Republican will be expected to move in. While politicians representing the British Labour Party or Brazil's Workers' Party are expected to adhere to a platform in order to receive the party's endorsement, politicians in the United States can call themselves Democrats or Republicans as long as they win their party's primary election.

That's why primaries are critical in influencing the direction of both major parties: as renowned political scientist E. E. Schattschneider said, "He who can make the nominations is the owner of the party." Primaries are the venue in which members of the Democratic Party get to debate and decide who can receive the official nomination and call themselves "the Democrat." Once you're a "Democratic nominee for office," your policies and proposals are part of the party.

Primaries also provide another way to leverage the people power that movements build. Movements push for change from the outside and are often focused on changing public opinion and forcing an issue onto the national agenda; then parties translate those changes into law. But parties can choose to ignore the protests and public opinion or dilute the movement's demands with compromises.

But if the movement's demands are popular with the base of the party, and now the politicians who chose inaction are facing primary challengers, suddenly the politicians might start singing a different tune. And if a few of those politicians get ousted by movement-aligned challengers, suddenly there are champions in the halls of power ready to translate the public sentiment and protest of movements into law.

These champions who run in primaries don't only give the movement power inside Congress; they also expand the base of the movement itself. Think about how many more thousands of people participated in Bernie Sanders's campaign than participated in Occupy Wall Street. Primaries can also lend movements more political force. Elections uniquely allow candidates to weave the multiple constituencies and demands of social movements into a governing vision that is greater than the sum of its parts. And by putting that vision on the ballot, candidates "politicize" the demands of movements and create consequences—losing a primary election—for opposing the movement.

My experience of Dave Brat's primary upset over Eric Cantor taught me that the candidate who can unseat an incumbent in his or her own party ends up receiving an enormous amount of influence over *articulating the direction* of the Democratic Party. With Alexandria Ocasio-Cortez's victory over Rep. Joe Crowley, that's exactly what happened.

I first met AOC a few weeks after joining Justice Democrats, at a Thai restaurant a few blocks away from the restaurant where she was bartending. AOC was challenging Crowley, a party veteran widely reported to be the next Speaker of the House after Nancy Pelosi. She knew that it was going to be hard, probably impossible, but she was willing to give it a shot. Out of all the candidates that the organization had recruited to run, I felt that she had the best shot of actually winning and creating the kind of earthquake that the Tea Party had created by unseating Eric Cantor.

While Crowley wasn't a Republican-in-disguise who voted with the GOP on issues like guns or abortion, he was a white male ten-term incumbent who received millions of dollars from big corporate donors on Wall Street despite representing a diverse, working-class district. Crowley weakened financial regulations in the Dodd-Frank bill by introducing provisions directly written by the big banks. He accepted campaign contributions from the same people bankrupting Puerto Rico, and he had little to say about mass deportation and the growing grassroots demands to #AbolishICE. Despite representing a district that voted for Barack Obama over Mitt Romney by nearly 90 points, Crowley had never led on any of the issues electrifying progressives in the era of millennial movements and Bernie Sanders's campaign. Throughout the campaign, AOC hammered Crowley with the motto of Justice Democrats: *It's time for a Democratic Party that fights for its voters, not corporate donors.*

"I'm not running 'from the left,' I'm running from the bottom," AOC wrote on Twitter shortly after defeating her opponent.

While Sanders represented many of the movements pushing the Democratic Party ideologically, AOC also represented them demographically. She was a millennial woman of color and skilled at communicating a populist call for multiracial democracy in one of the most diverse districts in America.

Within a few months after her victory, Ocasio-Cortez had teamed up with Justice Democrats and the Sunrise Movement to launch the campaign for the Green New Deal outside Nancy Pelosi's office. And because she, like Dave Brat, had both unseated a longtime incumbent Democrat and was picking an open fight with the leadership of her party, Ocasio-Cortez and the movement she aligned herself with received a platform to talk about what direction the Democratic Party needed to go.

For a generation in a hurry, the hypothesis that primary challenges could make change fast has been proven correct.

Unseating incumbents is the clearest way our movements can demonstrate our political power in terms of sheer votes—and occasionally elect a game-changing champion to Congress. Increasingly in primary elections across the country, younger progressive insurgents aligned with social movements have drawn lines against incumbents aligned with big corporate donors and establishment networks of patronage.

AOC's campaign against Crowley wasn't a fluke. Our movements were joining the fight over the soul of the Democratic Party.

In March 2016, Kim Foxx defeated Anita Alvarez in Cook County, Illinois, by aligning herself with the Black Lives Matter movement. In early 2017, Tom Perriello challenged Ralph Northam and Virginia's Democratic Party establishment from the left in an ultimately unsuccessful bid for governor. Larry Krasner (Philadelphia), Rachael Rollins (Suffolk County, Massachusetts), and Chesa Boudin (San Francisco) became the most progressive district attorneys in America, often by working alongside organizers from social movements and the Sanders campaign.

And throughout the 2018 primary season, grassroots, progressive candidates took on party-backed candidates. Some were victorious, most notably Ocasio-Cortez, Ayanna Pressley, Ilhan Omar, and Rashida Tlaib—otherwise known as "the Squad." Others, like Tiffany Cabán and Abdul El-Sayed, were unsuccessful but galvanized a network of loyal supporters and pushed the party in a progressive direction on criminal justice reform and health-care policy, respectively.

FACTIONS DRIVE THE BATTLE FOR THE SOUL OF THE PARTY

There's a fight for the soul of the Democratic Party because American parties are so big. Different ideologies fit into these big tents, and factions inside the party need to fight it out to decide the direction of their agenda. In other countries with a multiparty system, you would have more than two political parties. Typically, the breakdown in a multiparty system looks something like this: a nativist party, a business-aligned conservative party, a liberal party, a social democratic party, and a democratic socialist party. But in the American two-party system, these competing ideologies are all collapsed within the broad coalitions that make up the Republican and Democratic parties. This means that the Republican Party coalition contains factions of neoconservative hawks, free marketeers, the religious Right, and Trumpist xenophobes.

What's emerging in today's Democratic Party is an ideological faction that is aligned with both social movements and the increasingly young, multiracial base of the party, and its flexing its muscle in primary fights across the country. The other faction? The Democratic Party establishment—some of whom have no core beliefs and just follow the directives of big donors and party leaders, and others of whom adopted Third Way Clintonism during the Reagan era.

As political scientist Daniel DiSalvo argues, factions are the "engines of change that develop new ideas, refine them into workable policies, and move them through the halls of government."

American political parties are "big tents" that often lack a clear, singular position on any given issue. That's why it's nearly impossible to answer the question: *What is the Democratic Party's policy on climate change? Health care? Tackling inequality?* If you ask the establishment, you'll typically get vague answers focusing on "values" and "access to opportunities" alongside criticisms of the Republican Party from the highest levels of Democratic Party leadership. If you ask the rising progressive faction, you'll hear something different—the

Green New Deal, Medicare for All, a wealth tax, free public college and universities, and more.

Democrats are hungry for a party that campaigns not just on what we're against but on what we stand for. The Wild West openness of America's political system gives factions an opportunity to significantly tilt the direction of their party if they're united around a clear vision and willing to pick some fights.

In a political system where presidents and party leaders are incentivized to suppress their policy positions so that they can bargain with "centrists" in order to achieve majorities necessary to pass legislation, factions always have leverage—*if they choose to use it.*

PUTTING IT ALL TOGETHER: PRIMARIES AND THE INSIDE-OUTSIDE STRATEGY

While it may not be noticeable to most Americans, social movements aligned with ideological factions determine the direction of our political parties. Factions shape the direction of a party through primary fights—which determine what the party is and what it stands for. Political scientist Daniel Schlozman argues that successful social movements "anchor" parties: "They deploy resources to favored candidates, forcing politicians to respond to these alternative bases of authorities . . . [they are] polarizers with troops."

If successful, it works like this:

1. Movement Begins.
   Sunrise Movement launches, as many Democratic voters have become disgruntled with their party's positions on the economy and climate change, particularly after Occupy Wall Street and the fights around the Keystone and Dakota Access pipelines.

2. Movement Brings Attention to Issue.
   Through organizing and protest, Sunrise brings greater attention to the climate crisis and amplifies stories of millennials who are watching elected officials rob them of a safe, livable future.

3. A Faction Takes Up the Movement's Cause—by Winning a Primary Challenge.
   Justice Democrats recruit and elect Ocasio-Cortez, who beats Rep. Crowley. A new faction grows in the Democratic Party. Ocasio-Cortez teams up with Sunrise at the November sit-in and later introduces a resolution, pushing Nancy Pelosi and Democratic Party leaders to respond to a new force and new ideas in the party.

4. Movement's Issue Becomes Party Priority.
   Sunrise and Justice Democrats pressure Democrats to support the Green New Deal. Many incumbents are afraid of primary challenges and don't want to feel like they're out of step with the party's base. Climate change ascends as a top-three concern among Democratic primary voters. Nearly all the major Democratic presidential contenders say they support the Green New Deal, and over one hundred members of Congress have co-sponsored the resolution.

5. We Win.
   Factions energize new constituencies of voters that help defeat Republicans to win Congress and the White House. The movement prepares to pressure the party to force the issue into legislative action. See you in 2021!

This isn't a new strategy.

Factions have emerged throughout our country's history in

times of political or economic crisis. The Freedom Caucus aligned with the Tea Party's xenophobia and pushed the Republican Party to play constitutional hardball against President Obama. The New Democrats of the 1980s and 1990s, headquartered in Bill Clinton's Democratic Leadership Council, moved the Democratic Party away from "big government" New Deal and Great Society programs and toward more "market-friendly" ideas like cutting welfare programs and deregulating Wall Street. In the 1970s, evangelicals moved the Republican Party toward unabashed social conservatism. In the 1930s, the Congress of Industrial Organizations allied with staunch New Dealers in Congress, such as Senator Robert Wagner, who pushed their party to strengthen labor unions and invest in public infrastructure projects.

The most famous faction in American history was the Radical Republicans. Aligned with the growing abolitionist movement in America, they sought to abolish slavery and reconstruct the South, while President Lincoln and many Republicans were more inclined toward a moderate approach.

They discovered not only that their faction was unable to succeed without compromising with President Lincoln's incrementalism but that it was their responsibility to point out the contradictions within Lincoln's own approach to emancipation and seek policy and personnel concessions from his administration.

When Lincoln proposed a bipartisan bill that would provide federal compensation to the governments of border states in exchange for the gradual abolition of slavery, Thaddeus Stevens—an outspoken Radical Republican leader—retorted, "It is about the most diluted, milk and water gruel proposition that was ever given to the American nation."

The Radicals "are nearer to me than the other side, in thought and sentiment, though bitterly hostile personally," Lincoln said in

1863. "They are utterly lawless—the unhandiest devils in the world to deal with—but after all their faces are set Zionwards."

While the faction of Radical Republicans wasn't always unified, this group of legislators often understood their role and came together around shared goals.

Radicals flexed their political power in primary contests within the Republican Party. When radical abolitionist Owen Lovejoy defeated a moderate Republican in an Illinois party primary, Lincoln remarked, "It turned me blind . . . seeing the people there, their great enthusiasm for Lovejoy, considering the activity they will carry into the [general election] contest with him."

When factions rise, it's important to pay attention to how people outside the faction react. There are people like Bernie Sanders and the Squad, who represent the faction in its purest ideological form. But there are also people like Senator Ed Markey, party veterans with seniority who feel the winds changing and choose to side with the growing movement at the base of the party. Leaders like Markey know that the solutions that match the scale of the climate crisis aren't winnable unless there is activist energy demanding more from their own party, expanding the bounds of what's considered possible.

As with the Radical Republicans of the 1850s and 1860s and the New Right in the 1970s, the progressive faction today aspires to define the next generation of American politics and win a majority to its cause. "They are wisely acting as if they represent the demographic and political majority that their generation will become," historian Barbara Ransby wrote about the Squad. "They are not only the future of the Democratic Party. They are the future."

They are offering Democrats a multiracial, social democratic vision through their advocacy of the Green New Deal and other solutions that match the scale of the crises we face. For too long, the Democratic Party thought they could win by moving toward

Republicans and giving their voters half measures that were approved by their corporate donors. Now, Democrats are learning that the way to win is by fighting for policies that take money and power from the elite and give it to working people of all backgrounds.

To win, we will need to hold fast to that vision of an America for all of us.

A CODA: PASSING A GREEN NEW DEAL

We're not naive. It's clear that simply electing more Justice Democrats like Alexandria Ocasio-Cortez to Congress won't be a silver bullet in helping pass a Green New Deal. It's even more evident that Republicans won't come to some epiphany and decide to work with Democrats on climate legislation. We need major structural reforms to our democracy in order to break the gridlock in order to create a government that can actually pass a Green New Deal.

Some might argue that by electing more leaders like the Squad to Congress, we're creating more polarization and gridlock. That's wrong. First, the polarization already exists between our political parties. It began with Republicans. Members of the Squad are simply providing a countervailing force. They aren't afraid to highlight the polarization for what it is, two fundamentally different visions for our country, an America for the few or an America for all of us. Second, parties are supposed to stand for a clear vision. In most other countries, parties usually aren't expected to "reach across the aisle" and whittle down their values to pass something that activists in both parties despise.

The divisions in our country aren't going to go away, and Green New Dealers are going to have to jump through the Overton window one day.

SIXTEEN

REVIVING LABOR, IN NEW DEALS OLD AND GREEN

BOB MASTER

After some forty years of ideological and legislative abuse and neglect, the idea of the New Deal has been unexpectedly revived by the burgeoning youth-led climate movement. The call for a Green New Deal invokes memories of the original New Deal's unprecedented deployment of federal resources and power to revive the economy, rein in unfettered corporate power, and begin weaving a social safety net during the worst economic crisis in the nation's history. This development ought to be welcomed with open arms by the American labor movement.

After all, it was the 1930s New Deal that gave birth to the modern labor movement, when a handful of insurgent unions broke from the conservative American Federation of Labor to form the Congress of Industrial Organizations, signed up millions of auto, steel, rubber, and electrical workers, and transformed labor relations in the industrial heart of America's economy. It was the moment when, after seventy years of often militarized hostility, the American government at last put its thumb, however tentatively, on the workers' side of the scale and decreed, in the words of the National

Labor Relations Act, that the "practice and procedure of collective bargaining" and the workers' right to "full freedom of association [and] self-organization" would be "the policy of the United States."

But more than nostalgia for the golden era of industrial organization should inspire labor to embrace the Green New Deal. We need a Green New Deal, first and foremost, because with each 500-year flood, each catastrophic fire season, every instance of coastal flooding or arctic ice shear, the terrifying immediacy of the climate crisis has become increasingly clear. The primary mission of workers and their unions is to wage the unending battle for economic justice and dignity in the workplace. But we—and our children and grandchildren—need a habitable planet on which to carry on that struggle, and the Green New Deal, with its ambitious proposals for a thorough transformation of the US economy, is singular in its call for a collective response commensurate with the urgent crisis we confront. The Green New Deal reminds us that our nation has grappled with massive social and economic crises in the past and that we can do it again, if we are willing to shed ideological constraints and undertake the massive social mobilization needed to preserve the planet.

But there is another critical, though perhaps less apparent, reason for labor to embrace the Green New Deal, and it lies in a lesson drawn from the original New Deal as well. The New Deal of the 1930s represented not only an abandonment of past habits of governmental inaction in the face of economic crises, along with a vast expansion of the role played by the federal government in addressing the day-to-day needs of ordinary Americans. It also embodied a decisive ideological shift, a break from the Social Darwinist, laissez-faire philosophy that, but for a brief interlude during the Progressive Era, marked the limits of government engagement in American society since the end of Reconstruction. During the New Deal, the very idea of American freedom was transformed, as "freedom from want" eclipsed "freedom of contract," and a new regime of "industrial democracy" was counterposed to the ruthless entrepreneurial individualism of the Gilded Age. This shift played an indispensable role in seeding the soil of industrial unionism; it provided the emergent labor movement with the ideological rationale and confidence to mount its audacious challenge to the barons of industry. The new labor movement was so closely synchronized with the ethos of the New Deal that historian Michael Denning has labeled the era "the age of the CIO"; the " 'industrial unionism' " of the 1930s, he writes, "was not simply a kind of unionism, but a vision of social reconstruction."

Today's labor movement desperately needs a comparable revolution of ideas and values; despite potentially significant signs of revival in recent years, two generations of ideological battering have taken a heavy toll, contributing to labor's long-term marginalization and delegitimization. Here is the second crucial rationale for labor to support the Green New Deal. The GND is not just a framework for climate action. It is also a needed intervention in the nation's political discourse, an attempt to provoke an ideological U-turn on the scale of what took place eighty years ago. The Green New Deal calls for building a new, green economy in which unionized workers will play a central role, and where the interests of workers and communities will take priority over the greed and self-interest of the 1 percent.

To their great credit, this vision reflects the lengths to which the young organizers behind the Green New Deal have gone to build a bridge to labor. The GND resolution places union workers at the forefront of a green economic future and proposes to guarantee wage and benefit parity to workers who transition out of high-carbon sectors. It explicitly makes workers' rights central to any plan to address the climate crisis and commits to "ensuring that the Green New Deal mobilization creates high-quality union jobs that pay prevailing wages, hires local workers, offers training and advancement opportunities, and guarantees wage and benefit parity for workers affected by the transition." And it states that "strengthening and protecting the right of all workers to organize, unionize, and collectively bargain free of coercion, intimidation, and harassment" will be a central objective of any climate mobilization. Green New Dealers know that the climate crisis is just one consequence of a broader corporate offensive that has produced profound crises of racial and economic justice as well. Deregulation, free trade, slashing social programs, tax cuts for the wealthy, and the attack on workers and their unions have been the building blocks of a comprehensive ruling class agenda since the mid-1970s. Yes, these policies endanger the planet, according to Green New Deal proponents, but they are also inseparable from the assault on poor and working people and their unions.

As a union staffer and activist for more than four decades, most of that time in the leadership of the northeast region of the Communications Workers of America, I have had a front-row seat to corporate America's escalating attack on workers. President Reagan's firing of over 11,000 striking federal air traffic controllers in 1981 heralded an era of private sector strike-breaking that left our movement littered with defeated strikes and broken unions. The bargaining and organizing rights of workers have been steadily eroded by hostile courts and Republican appointees to the National Labor Relations Board. Conservative legal activists spent decades successfully plotting to undermine public sector unionism, and Republican governors put dismantling public sector bargaining rights at the top of their agendas after the 2010 election cycle. The Democratic Party was, at best, ambivalent about these developments. In power, it repeatedly failed to prioritize strengthening labor law, sacrificing the opportunity to stanch the hemorrhaging of labor's ranks, not to mention preserve a crucial section of the party's electoral base. And its succession of neoliberal presidents, from Carter to Clinton to Obama, supported "race-to-the-bottom" free trade agreements, as well as the deregulation of industries like telecommunications, transportation, and airlines, which severely undermined union power. Campaign season rhetoric about the needs of working men and women rarely translated into pro-labor governance.

The economic consequences of this "one-sided class war," as UAW president Doug Fraser dubbed it back in 1978, are by now familiar: stagnant incomes, soaring inequality, crushing levels of debt, persistent racial and gender disparities in income and opportunity, and the proliferation of precarious employment. These trends are both symptom and cause of the decline of private sector union density to 6.4 percent, a level not seen since the decade prior to World War I.

This forty-year assault on unions was also embedded in a sophisticated intellectual project, a multifaceted campaign to destroy the ideological foundations of the New Deal. In 1971, a Virginia-based corporate lawyer named Lewis Powell, soon to be nominated to the Supreme Court by President Nixon, wrote a now infamous memo for the US Chamber of Commerce, outlining how the business community should respond to what was perceived as the dire threat posed by the anticorporate New Left. Warning that the "survival of what we call the free enterprise system" was at stake, Powell's memo offered a blueprint for corporate America's plan to transform political discourse. Think tanks, university faculties, coordinated legislative initiatives, and the mass media would all be weaponized against the aging New Deal consensus. Corporate mouthpieces like *Fortune*, *Business Week*, and the *Wall Street Journal* pinned US economic malaise in the 1970s and 1980s on restrictive union work rules and overly generous contractual wages and benefits.

These attacks on labor complemented hand-wringing over the "middle-class" tax burden and incentive-killing "death taxes," as well as denunciations of Cadillac-driving "welfare queens," and "law and order" appeals to the "silent majority." "White supremacy" and "property supremacy" went hand in hand in the corporate offensive against the New Deal, historian Nancy MacLean has pointed out, and as Ian Haney López describes elsewhere in this volume, the dog whistle racism of the New Right served to distract working-class attention from corporate attacks on labor and the social programs upon which working people depend. In the decades of Reagan, Thatcher, Clinton, and Blair, neoliberalism grew into "the most successful ideology in world history," as one observer put it, and the exaltation of markets and individual enterprise left little space in the public imagination for unions.

Amid this wreckage, labor has debated strategies for reviving the movement and reversing its decline in membership for the better part of two decades. Any revival will require a greater willingness to take risks, break laws, walk picket lines, and innovate new organizing strategies. But ultimately the labor movement is as much a political and ideological project as an organizational one. It flourishes when its mission resonates with broader aspirations stirring in society, when worker organization helps to answer fundamental social and political questions that transcend any individual workplace. For example, the explosion of public sector organizing in the 1960s channeled the civil rights consciousness of the Black Freedom and women's movements into government workplaces—where a heavily of-color and female workforce demanded access to the "middle-class" living standards and on-the-job protections that (mostly) white male workers had won in the 1940s and 1950s.

Labor's revival requires a shift in the dominant ideas of politics and the economy, a shift that endows unions with social, political, and moral authority and purpose. It is in this context that we can begin to appreciate the potential significance of the Green New Deal to the fortunes of the labor movement. The GND presents a generational opportunity for labor to participate in building a movement that can put an end to the dominance of anti-labor ideas and right-wing coalitions, and to assume a leading role in building a green economy that puts workers and their unions at the center of national life once again. The climate movement is the future of grassroots insurgency; it promises to grow in size and influence as the crisis intensifies. Labor can restore its own sense of mission and relevancy by joining arms with climate activists.

The history of the New Deal demonstrates that ideas matter, and that movements do not triumph on the strength of militancy alone. It took a complex interplay of worker insurgency, shifting hegemonic views of politics and economics, and the leveraging of government power to enable labor to win the establishment of permanent industrial unions. The explosion of militancy among rank-and-file workers across the economy in response to the economic devastation of the early Depression years was unarguably the indispensable precondition for the rise of the CIO. From Gastonia, North Carolina, to Biddeford, Maine, from the coalfields of eastern Kentucky to the docks of San Francisco, in industrial centers like Toledo, Akron, Detroit, Chicago, and Cleveland, fury at wage cuts and mass layoffs ignited militant confrontations between workers and authorities. But the militancy of the early Depression years alone did not suffice to win the establishment of permanent, powerful unions across industry.

Industrial insurgency did throw fear into the hearts of corporate elites and opinion makers, and created pressure on Congress to act to contain the growing workplace conflict. And the 1934 election produced lopsided congressional majorities for the Democrats, setting the stage for the New Deal's progressive high-water mark of 1935. The Congress elected in 1934 was perhaps the most radical in American history, before or since. "This is our hour," said Harry Hopkins, one of the most progressive members of FDR's "brain trust." "We've got to get everything we want—a works program, social security, wages and hours, everything—now or never."

The National Labor Relations Act, sponsored by New York senator Robert F. Wagner, was a central plank in the remarkable legislative agenda of 1935, the so-called Second Hundred Days, along with Social Security, rural electrification, progressive taxation, and the antimonopoly Public Utility Holding Company Act. Later developments—the passage of the Taft-Hartley Act in 1947, restrictive court rulings, as well as the emergence of a vicious union-busting industry in the 1970s and 1980s—would eventually hollow out the NLRA, leaving it a flimsy scaffold for the protection of workers' organizing rights. But at the moment of passage, the enshrinement of the workers' right to self-organization, and the imposition on American capitalists of a statutory duty to engage in collective bargaining, left the business community aghast and enraged. One business publication labeled the act "one of the most objectionable, as well as one of the most revolutionary pieces of legislation ever presented to Congress," and an Oklahoma business association objected that the Wagner Act would "out-SOVIET the Russian Soviets." "When passed," legal scholar Karl Klare has written, "the National Labor Relations (Wagner) Act was perhaps the most radical piece of legislation ever enacted by the United States Congress."

But even this stunning legislative achievement did not overnight translate into the massive organizing campaigns we associate with the rise of the CIO. As historian Steve Fraser has written, it took

FDR's landslide victory in the 1936 election, one of the most class-polarized contests in the country's history, to "unleash a mass movement of unprecedented militance and tactical boldness." Fraser, along with other leading New Left historians of the CIO like Nelson Lichtenstein, Ronald Schatz, and Peter Friedlander, have compellingly argued that it was the cultural and political transformation of millions of first- and second-generation semiskilled industrial workers during the historic election campaign of 1936 that ultimately endowed the industrial working class with the confidence to challenge the corporate tyranny of basic industry. This analysis adds important texture to an often simplified account of how rank-and-file militancy and mass strikes alone provided the keys to labor's resurgence.

As Roosevelt barnstormed the Midwest in the fall of 1936, hundreds of thousands of autoworkers—mostly southern and eastern European immigrants, and none of them yet members of the fledgling UAW—rallied in auto manufacturing centers like Flint, Pontiac, and Detroit to cheer on the president. As Lichtenstein writes, "It slowly dawned on activists like [Walter] Reuther that the pro-Roosevelt excitement sweeping the working-class wards of the city might provide the key to the breakthrough that had so long eluded them . . . The 1936 election mobilized the ethnic working class as no other, decisively confirming the power of the emergent Roosevelt coalition." After the president's triumphant victory, the CIO moved quickly to transmit this sense of empowerment from the ballot box to the workplace. " 'You voted New Deal at the polls and defeated the Auto Barons,' union organizers [had] told Michigan workers late in 1936. 'Now get a New Deal in the shop.' "

The emergence of the CIO was the result of multiple intersecting developments. Rank-and-file militancy disrupted production, created unsustainable tension in workplaces across the country, terrified elites, and generated irresistible pressure on Congress to address the industrial crisis. The Wagner Act provided government sanction for workers' bargaining and organizing rights. Left-wing activists, communists and socialists of every description, worked tirelessly to educate workers, build shop-floor committees, and plot strike strategy. And the watershed election of 1936 shifted the consciousness of millions of workers and created an unprecedented sense that the leadership of the US government stood on the side of workers. "As one millworker in the South noted," according to Lichtenstein, "Roosevelt 'is the first man in the White House to understand that my boss is a son of a bitch.'" These developments coalesced to produce the enormous achievements of the CIO—"one of the greatest chapters in the historic struggle for human liberties in this country," in the words of David Montgomery.

This account of the New Deal suggests lessons for Green New Dealers and labor activists alike. First, we can clearly see the importance of ideas and ideological shifts in the chain of events that set the stage for the triumph of the CIO. In just a few short years, longstanding dogma about limited government and individual responsibility was replaced by a vision of expansive government rooted in a newfound sense of social solidarity. The Green New Deal is contributing to an analogous process today, with its explicit emphasis on workers' rights, and by merging the agendas of climate, racial, and workplace justice. By placing the questions of labor and economic inequality at the center of their vision for a new, green economy, proponents of the Green New Deal offer the labor movement a potential sense of social mission that is necessary for its revival.

The GND is by no means the only factor prying open this ideological space, as the *New York Times* noted in mid-October 2019. In the early months of the 2020 Democratic primary campaign, most of the leading candidates had embraced far-reaching pro-labor platforms that include strengthening organizing rights, restricting the definition of independent contractors, and restoring union rights to engage in secondary boycotts. These shifts reflect the growing acknowledgment in academic and elite circles that the decline of unionism has been a major contributor to soaring income inequality over the last forty years. "The increased openness to unions and collective bargaining . . . reflects a broader ideological shift in the country away from the market-friendly policy approach of the 1980s and '90s," wrote *Times* correspondent Noam Scheiber, "which has lost credibility as inequality has widened."

The decade following the financial crisis of 2008 gave rise to two important social movements, Occupy Wall Street and Black Lives Matter. It also saw political breakthroughs such as the Bernie Sanders campaign in 2016, the Sanders and Elizabeth Warren campaigns in 2020, and numerous state and local campaigns that explicitly challenged a bankrupt neoliberal consensus. This shifting political discourse has already had a beneficial impact on the labor movement. In 2016, when over 36,000 workers struck against Verizon for seven weeks, Sanders was omnipresent on the picket line, and he denounced the company's CEO during the nationally televised New York Democratic presidential primary debate. The years 2018 and 2019 saw a sharp uptick in strikes, putting more workers on the picket line than at any time in the last thirty years. There has been an upsurge in organizing activity, particularly among young workers, at newspapers, in digital media, and, more recently, among digital game programmers. In 2019, both the United Teachers of Los Angeles and the Chicago Teachers Union waged remarkable strikes "for the common good," demanding not just raises and smaller class sizes but improved school-based social services, limits on charter schools, and, in Chicago's case, increased investment in housing for both students and teachers.

It is premature to predict a decisive turnaround in labor's fortunes or organizing upsurges on the scale of the 1930s and 1940s. But there is no doubt that significant leftward shifts in the ideological climate have provided a shot in the arm to the labor movement, not nearly on the scale of the Wagner Act in 1935 and Roosevelt's reelection of 1936 but significant nonetheless.

A second lesson from the New Deal era for both climate and labor activists is the decisive role of politics and the government. Agitation and mass mobilization are undoubtedly critical in shifting the boundaries of what is considered governmentally possible in any struggle for structural reform. We have seen this again and again in struggles for social justice, from abolitionism to the labor movement, from the Black Freedom struggle of the 1960s to Black Lives Matter, from the Fight for $15 to the climate justice movement, from gay liberation to the standoff at Standing Rock. But the corollary is also true: absent governing power, victory has historically remained beyond the reach of social movements.

When the abolitionists launched their campaign to end slavery in the mid-1830s, most were moral crusaders who abjured the very act of voting as complicity with a constitutionally sanctioned slave system. But ultimately, leading abolitionists like Frederick Douglass, Wendell Phillips, and William Lloyd Garrison needed Radical Republican legislators like Thaddeus Stevens, Charles Sumner, and Salmon Chase—and ultimately President Abraham Lincoln—to complete the crusade to end slavery. Likewise, the CIO needed Senator Wagner and his congressional allies to create a new framework for collective bargaining, and Martin Luther King Jr. needed President Johnson to wield his legendary legislative prowess to win passage of the Civil Rights, Voting Rights, and Fair Housing Acts of the mid-1960s.

Demonstrations alone will not produce the regulatory changes needed to stem the climate crisis. We must win governing power.

That is the breakthrough significance of the Green New Deal resolution—an effort to translate the agitation of the street into a public policy initiative equal to the ambition of the original New Deal.

The Green New Deal resolution has drawn fire from some quarters for being overly vague and unrealistically ambitious. But a third lesson from the 1930s is that the New Deal did not spring forth upon FDR's inauguration as a fully realized progressive program to resurrect the shattered US economy. The policies of the New Deal were contingent and contradictory, scarred by unavoidable legislative compromises with the entrenched congressional power brokers of the Bourbon South. Roosevelt himself was no progressive ideologue. He campaigned in 1932 as a fiscal conservative, pledging to slash government spending by 25 percent. This wasn't mere campaign rhetoric, either. Just six days after his inauguration, he sent the Economy Act to Capitol Hill, authorizing a 15 percent cut in the salaries of federal workers and hundreds of millions of dollars more in cuts to pensions for veterans. It was adopted the next day. The legislation was the work of Roosevelt's budget director, Lewis Douglas, the fiercely anti-union and fiscally conservative nephew of the CEO of the Phelps Dodge mining company, Walter Douglas, who was notorious for having deported a trainload of over a thousand striking members of the Industrial Workers of the World into the searing Arizona desert in 1917.

Roosevelt was a pragmatist, surrounded by an eclectic set of advisers, who quickly turned to more progressive responses to the Depression crisis, including the Agricultural Adjustment Act, the Civilian Conservation Corps, the National Industrial Recovery Act, the Tennessee Valley Authority, the Federal Trade Commission, and the Glass-Steagall Act—the alphabet soup of agencies and initiatives that we associate today with the period referred to as the First

Hundred Days. But the contours of his program were always subject to struggle, both inside and outside the administration. It was the intensifying mobilization of workers, farmers, and senior citizens in 1934 that made possible America's social democratic moment in 1935, when the Wagner Act and the second New Deal passed. Likewise, the substance of any Green New Deal will be shaped by continuing struggle, in the streets, at the ballot box, and in the halls of the legislature. The powerful impact of growing youth movements on the climate debate is already apparent, but much, much more will be needed to define and bring into reality a Green New Deal.

Steve Fraser has written that in the first decades of the twentieth century, the "labor question"—the fortunes of a growing and increasingly exploited working class in the "new industrial order"—was the "constitutive moral, political, and social dilemma" of the era. The New Deal of the 1930s provided the capitalist system's answer to that question. Like the "labor question" during the first third of the twentieth century, the climate crisis is the defining issue of *our* time. The labor movement cannot stand aloof from the fight to preserve the planet. Indeed, a labor movement that fails to take up the question of climate risks consigning itself to social irrelevancy.

Here the appropriate New Deal analogy may be the behavior of the American Federation of Labor during the 1930s. The leadership of the AFL failed to grasp the changing mind-set of many American workers and the upending of the class system in the Depression era. They disdained efforts to organize immigrant workers and fiercely opposed the enactment of the Wagner Act. Most AFL unions, representatives of skilled workers who usually exerted tight control over both work practices and hiring, had long practiced racial exclusion. Access to the skilled trades was determined by who one knew, and nepotism and racism defined the boundaries of opportunity. Indeed, as late as the mid-1940s, seventeen out of a hundred AFL affiliates and independent unions "still had a [constitutional] clause expressly limiting membership to 'whites' or 'Caucasians.' "

As a consequence, the AFL was marginal to the emergence of the transformational, progressive array of ideas and institutions that arose during the New Deal era. Today's labor movement risks a similar marginalization if we act only as a defensive veto player, narrowly focused on the interests of short-term job protection, and against popular initiatives like the Green New Deal. There must be a green transition, and it must account for those whose jobs in the carbon economy are at risk. The key is to focus on the tens of millions of new jobs—in green manufacturing, solar panel installation, and building retrofits—that will be created.

At the same time, the labor movement must recognize that young people are among the most pro-union in generations. They represent the best hope for sustaining a progressive movement that can create the conditions for a revival of the labor movement. They are also existentially terrified by climate change. If the labor movement fails to seize the opportunity opened up by the GND, we risk permanent marginalization and irrelevance within an emergent Left coalition, something that may doom both labor and the Left.

Like the Great Depression, the climate crisis demands a sweeping mobilization of all the resources and ingenuity at our disposal. The Green New Deal invokes the nation's experience of such a mobilization in the 1930s. Flawed and limited as it may have been, the New Deal nevertheless transformed the role of government in American society, dramatically expanded the scope of social provision, and produced the country's first permanent industrial labor movement. Central to its success was the establishment of workers' legal right to organize and bargain collectively.

The Green New Deal has made workers' rights, along with racial justice, integral to the social transformation that will be needed to contain and reverse the effects of climate change. In so doing, it signifies both an ideological repudiation of the forty-year attack on working people and a practical call for a massive campaign against climate change. It augments ideological changes already underway that recenter the importance of building working-class power in society. It holds the promise of contributing to the growth of a broad progressive movement that can create the conditions under which the labor movement may rise again. For all these reasons, labor must embrace the Green New Deal.

ORGANIZE. VOTE. STRIKE.

VARSHINI PRAKASH

When I was a kid, I set out to repair the harm being done to our planet. Fourteen-year-old Varshini believed recycling could save the world, so I joined my school recycling club. It didn't seem to add up, so I got involved with social movements, joining with hundreds and thousands more to demand change with a united voice. But the changes we won weren't the changes we needed to prevent a full-blown crisis. We realized we had to erode the economic power of the fossil fuel industry, build the might of our movements, and take the fight into the political arena. I grew up with a strong distaste for politics, wanting nothing to do with it. But once I came to terms with the enormity of the crisis before us, I understood. For humankind to survive, we need nothing less than a once-in-a-century upheaval of our political, social, and economic institutions.

So here you are, with a book about climate change, social movements, social justice, economics, and politics. When trying to figure out one, we have to look at the whole.

If I'm being truthful, we still haven't figured it out. Movements made leaps and bounds in the 2010s, but the opposition is still in power, and we're still on the outside. But there are a few things I'm sure of, lessons that I'm glad to be able to lift up through the incredible contributors to this book.

First: The climate crisis is worse than you think, and the Green New Deal is the commonsense solution. Tepid price signals or changes to your individual lifestyle will never get the job done. Now is the time for bold federal action—a guaranteed job for every person, to power the transition to renewable energy, sustainable agriculture, and low-carbon transit. The federal government can directly coordinate industrial activity while punishing corporations that shirk their responsibilities, just as it would in a wartime scenario.

Second: Our aim must be *governing power.* Not just the presidency, not just Congress, not just statehouses and city halls across the country, but all of the above, plus independent social movements to keep pushing the envelope and hold officials accountable. Power is *the ability to act*, and only with true governing power can we enact a Green New Deal.

Third: The more success we have as movements, the more those with money and power will attempt to stop us. They'll use every tool they have to divide our ranks and pit us against each other, because we're weaker apart than we are together. The greatest fault lines among us are race and class. When we decide that what unites us is more than what divides, and we make powerful efforts to include one another with integrity and true solidarity across differences, we take away the most effective weapon of the fossil fuel billionaires and the Donald Trumps of the world: divide and conquer. We have to build a multiracial, cross-class movement. It's the only way.

These are the things I know. But there's more that I don't know. I want you to pick up where I've left off and pick up the torch from here. Take this movement and run with it. Take these policies and design better ones. Learn from both success and failure to build more powerful campaigns.

The waters of our lives are choppier by the day, and the horizon full of storms. As we race to submit the final draft of this book, COVID-19 is claiming lives by the tens of thousands, sending people into social isolation to slow the spread of the virus, and making a mockery of my and everybody else's 2020 plans.

For this page, I had drafted a call for a massive youth uprising later this year, millions in the streets, to deliver an unmistakable mandate for the Green New Deal. That plan is now called into question, to say the least. While our theory of change—people power and political power—holds true, the details of how we get there are all under evaluation.

In the pandemic, we see laid bare the absolute necessity of competent government in times of crisis. We see the difference between a government using industrial policy to organize the production of ventilators and not doing so. The difference is measured in lives saved or lost. The same will be true—no, is already true—for climate change.

The pandemic demonstrates our fundamental interdependence. Looking out my window on Sumner Street, I see people dropping off food for neighbors under quarantine. I think of the nurses and doctors risking their lives daily, without question, for people they have never met. And I'm aware of the billions of people around the world who are making the strange sacrifice of staying away from others. At least for the moment, this is a time of unprecedented global cooperation. Maybe we can learn a few lessons to take with us after the outbreak is over.

As we were finishing the proofread of this book, we witnessed an uprising for black lives across America following the murders of George Floyd, Breonna Taylor, and countless other black Americans.

In the backlash to these historic mobilizations, we also saw disturbing, unjust violence wielded by police armed with weapons of war and funded with billions of dollars of public money. In their violence we see an America where justice and a livable future are not possible, where cries for freedom can be silenced by agents of the state and by a greedy few who cling to power and privilege on the basis of race.

But we've also seen an unprecedented demonstration of popular power and resistance. As protests spread across the nation, Derek Chauvin, the officer who killed George Floyd, was charged with second-degree murder, and a veto-proof majority of the Minneapolis City Council vowed to dismantle their police department and establish a new system of public safety. Again, just as history taught us, we see that if movements bring masses of people into intense, sustained action, we can win. These victories are the direct result of black communities engaging in collective struggle for decades.

These victories also affirm that a different future is possible. We are told so many things cannot be done. But we see billions invested into prisons, police, and weaponry—we see how government-constructed institutions of mass incarceration can rob communities of life and livelihood for decades. We know the question is not whether our country has the power to transform our society and our economy. The question is how will we wield the immense powers and resources of government, and whose lives will our laws serve and protect.

Things are changing so fast. By the time this even goes to print, these reflections may seem sorely outdated. But that's the point, actually. We cannot identify with certainty the turning points of history ahead of us, but if anything is certain, it is that we are not returning to the economy or the politics we had before 2020. I can't tell you how the story ends, any more than you can tell me. We have to live it, and shape it, together.

But I know, in that story, that the victors are the ones who refuse to let dreams be dampened. The victors are the ones who refuse to let hearts harden and visions evaporate and their fierce love for justice erode. In the immortal words of poet Aurora Levins Morales's "V'ahavta":

Don't waver. Don't let despair sink its sharp teeth
Into the throat with which you sing. Escalate your dreams.
Make them burn so fiercely that you can follow them down
any dark alleyway of history and not lose your way.
Make them burn clear as a starry drinking gourd
Over the grim fog of exhaustion, and keep walking.

Hold hands. Share water. Keep imagining.
So that we, and the children of our children's children
may live.

A new economy, a new world, a new way of being is possible, if each and every one of us escalates our dreams.

ACKNOWLEDGMENTS

This book is more than the product of the contributing authors and two editors. We'd like to thank those who made this project possible.

THE MOVEMENT. We wouldn't be editing chapters about a Green New Deal if it weren't for the thousands of young people who took a leap of faith and organized until the Green New Deal became a new center of gravity in US politics. Thank you to the members of the hundreds of Sunrise hubs who experiment and take risks each day to grow our movement in every corner of the country.

Like every other ambitious project our movement takes on, this book got done because of dedicated volunteers. Irene Henry refined the writing on movement strategy until it shone as bright as the North Star. Ben Gilvar-Parke distilled the beating heart of the stories shared in this book. Zea Marty led a team of bibliographers—Mikhaila Bishop, Victoria Hsieh, Laís Santoro, Carly Gray, Jessica Finkel, and Tara Benavides—to track down every citation. We are grateful for many more Green New Dealers who lent a helping hand, especially Sarah Abbott, Aggie Agreros, Lauren Black, Jeremy Brecher, Erin Bridges, Rebecca Conway, Sam Eilertsen, Karthik Ganapathy, Libby Gatti, Miles Goodrich, Naomi Hollard,

Aracely Jimenez, Emily LaShelle, Mattis Lehman, Ilana Master, Lauren Maunus, Emily Mayer, Greta Neubauer, Stevie O'Hanlon, Alex O'Keefe, Alice Oshima, Zina Precht-Rodriguez, Sam Quigley, Deirdre Shelly, Howie Stanger, Brian Stillwell, Jacob Surpin, Evan Weber, Courtney Wise, and Seth Woody.

Lastly, the ideas in this book were nourished by the progressive organizations across the Green New Deal movement, especially the Working Families Party, Climate Justice Alliance, Indigenous Environmental Network, Sierra Club, Indivisible, It Takes Roots, Center for Popular Democracy, People's Action, Greenpeace, US Climate Action Network, SEIU, NY Renews, Frontline Detroit, No Fossil Fuel Money coalition, Justice Democrats, and young climate strikers across the globe. To all those unnamed: we love you, we see you, we are honored to work alongside you.

THE BOOK TEAM. For shepherding us through the unfamiliar world of publishing with unerring advice, we are grateful beyond words for our agent, Anthony Arnove. He is a mensch. We cannot imagine sprinting through the hoops of making a book without Anthony's seemingly bottomless well of wisdom. Our editor, Eamon Dolan, took a chance on us as first-time writers and editors, solved the puzzles of an essay compilation when we could not, and patiently provided precious time and guidance when we stumbled. In each draft and proposal, no matter how rough, Eamon excavated essential ideas with peerless clarity and precision. His editorial skill is a gift to writers and to our movement, and we are honored to have worked with him. This book only became an actual, physical book because Tzipora Baitch diligently—and with immense attention to detail—followed every word and loose end from proposal to printing. Along the way, Janet Byrne heroically read every word, replaced the ill-chosen ones, and double-checked the facts.

When we burned the midnight oil, a few brave souls offered their talent and company to help us finish by dawn. Will Lawrence, Sunrise cofounder and lifelong organizer, is the man behind the jacket cover: he saw the contribution a book could have before anyone else; he launched the writing process and kept us above water, plugging holes and steering the ship through rough tides. Aaron Jorgensen-Briggs graced almost every chapter of this book with his talent for turning bad phrasing into good syntax, and he even dispelled the despair of a writer's block. Garrett Blad has inspired thousands of young people to use storytelling as a tool for movement-building, and this book has stories to share because of his mentorship. Marcela Mulholland never loses sight of the big picture of US politics, and nudged our writing to answer the big questions with humor and hope. Max Berger's political wisdom helped us make sense of "realignment," and his unconditional support of a tired Guido carried this book to the finish line.

Our parents—Ramaa, Prakash, Sergio, and Monica—hosted too many editorial meetings in their homes, piling us with delicious Indian and Italian cuisine. They have encouraged our passions and convictions at every turn. Additionally, Varshini could not have survived 2019 without the affirmation and Allen Iverson–themed pep talks of her fiancé, Filipe, and Guido could not have survived the book without Sara's generosity and love.

THE MENTORS AND FOREBEARS. We are indebted to countless organizers, historians, and writers who, for generations before us, made the road to the Green New Deal by walking. In lieu of a full accounting, here are those whose work directly influenced our organizing and writing these past few years. Betámia Coronel, Cristina DuQue, Cathy Kunkel, Katie McChesney, and Becca Rast were Sunrise visionaries before the movement was named Sunrise.

Momentum, Training for Change, the Ayni Institute, Relational Uprising, People's Action, and the Wildfire Project taught us how to build movements and what leadership means. When our strategies failed us, Mark and Paul Engler, Daniel Hunter, Yotam Marom, Carlos Saavedra, and Jonathan Matthew Smucker showed us a way forward. Ted Fertik kindly demystified the historiography of the New Deal. Movimiento Cosecha, IfNotNow, Dream Defenders, and United We Dream equipped us with the tools needed to seed a movement. 350.org gave crucial support when Sunrise was just a Google doc. Among our role models, we return again and again to the innovations and vision of Martin Luther King Jr., Ella Baker, and the Student Nonviolent Coordinating Committee. The writings of Michelle Alexander, Dr. Robert Bullard, Frances Fox Piven, Barbara and Karen Fields, Michael C. Dawson, and Corey Robin were essential sustenance as we grappled with questions of justice and democratic transformation in the United States.

We bear responsibility for any errors in the final text.

NOTES

INTRODUCTION: *THE ADULTS IN THE ROOM* / VARSHINI PRAKASH

vii *Over 1.4 million people*: Eliza Barclay and Brian Resnick, "How Big Was the Global Climate Strike? 4 Million People, Activists Estimate," *Vox*, September 22, 2019, https://www.vox.com/energy-and-environment/2019/9/20/20876143/climate-strike-2019-september-20-crowd-estimate.

ix *to a cheering crowd*: Anne Barnard and James Barron, "Climate Strike N.Y.C.: Young Crowds Demand Action, Welcome Greta Thunberg," *New York Times*, September 20, 2019, https://www.nytimes.com/2019/09/20/nyregion/climate-strike-nyc.html.

ix *That day, 7 million people*: Olivia Rosane, "7.6 Million Join Week of Global Climate Strikes," *EcoWatch*, September 30, 2019, https://www.ecowatch.com/global-climate-strikes-week-2640790405.html.

x *Four million more young people turned out to vote*: John Della Volpe, "Midterms Saw Historic Turnout by Young Voters," RealClearPolitics, November 8, 2018, https://www.realclearpolitics.com/articles/2018/11/08/midterms_saw_historic_turnout_by_young_voters__138591.html.

xii *twelve years to rapidly*: Jonathan Watts, "We Have 12 Years to Limit Climate Change Catastrophe, Warns UN," *The Guardian*, October 8, 2018, https://www.theguardian.com/environment/2018/oct/08/global-warming-must-not-exceed-15c-warns-landmark-un-report.

xiii *The last major* attempt: David Roberts, "The Green New Deal, Explained," *Vox*, March 30, 2019, https://www.vox.com/energy-and-environment/2018/12/21/18144138/green-new-deal-alexandria-ocasio-cortez.

xvi *Reconstruction era following the Civil War*: Blain Roberts and Ethan J. Kytle,

"When the South Was the Most Progressive Region in America," *The Atlantic*, January 17, 2018, https://www.theatlantic.com/politics/archive/2018/01/when-the-south-was-the-most-progressive-region-in-america/550442/.

xvi *New Deal era*: Olivia B. Waxman, "How FDR's New Deal Laid the Groundwork for the Green New Deal—in Good Ways and Bad," *Time*, February 8, 2019, https://time.com/5524723/green-new-deal-history/.

xviii *Greta Thunberg has said*: Greta Thunberg (@GretaThunberg), Twitter, October 21, 2018, 12:36 p.m., https://twitter.com/gretathunberg/status/1054048784844505098?lang=en.

CHAPTER 1: *THE CRISIS HERE AND NOW* / DAVID WALLACE-WELLS

3 *since the beginning of the Industrial Revolution*: V. Masson-Delmotte et al., eds., "Summary for Policymakers," in *Global Warming of 1.5°C: An IPCC Special Report on the Impacts of Global Warming of 1.5°C Above Pre-industrial Levels and Related Global Greenhouse Gas Emission Pathways, in the Context of Strengthening the Global Response to the Threat of Climate Change, Sustainable Development, and Efforts to Eradicate Poverty*," IPCC, 2018, https://www.ipcc.ch/sr15/chapter/spm/.

3 *all of recorded human history*: Jeremy S. Hoffman, Peter U. Clark, Andrew C. Parnell, and Feng He, "Regional and Global Sea-Surface Temperatures During the Last Interglaciation," *Science*, January 20, 2017, https://science.sciencemag.org/content/355/6322/276.

3 *we have already left behind*: Andrew Freedman, "The Last Time CO_2 Was This High, Humans Didn't Exist," Climate Central, May 3, 2013, https://www.climatecentral.org/news/the-last-time-co2-was-this-high-humans-didnt-exist-15938.

3 *wildfires in the American West has doubled*: "Climate Change Indicators: US Wildfires," WX Shift, Climate Central, https://wxshift.com/climate-change/climate-indicators/us-wildfires.

3 *large fires has quintupled*: "The Age of Western Wildfires," Climate Central, September, 2012, https://www.climatecentral.org/wgts/wildfires/Wildfires2012.pdf.

3 *60 percent of animal populations on earth have died since then*: Elizabeth Davis and Katie Walsh, "WWF Report Reveals Staggering Extent of Human Impact on Planet," World Wildlife Fund, October 29, 2018, https://www.worldwildlife.org/press-releases/wwf-report-reveals-staggering-extent-of-human-impact-on-planet.

4 *as have perhaps 70 percent of insects*: Mary Hoff, "As Insect Populations

Decline, Scientists Are Trying to Understand Why," *Ensia*, October 30, 2018, https://ensia.com/features/insects-decline-armageddon-biodiversity/.

4 *a quarter of their potential GDP*: Noah S. Diffenbaugh and Marshall Burke, "Global Warming Has Increased Global Economic Inequality," National Academy of Sciences, April 2019, https://www.researchgate.net/publication/332581715_Global_warming_has_increased_global_economic_inequality.

4 *just a few decades ago*: Fiona Harvey, "Greenland's Ice Sheet Melting Seven Times Faster Than in 1990s," *The Guardian*, December 10, 2019, https://www.theguardian.com/environment/2019/dec/10/greenland-ice-sheet-melting-seven-times-faster-than-in-1990s.

4 *three times in a single summer*: Matthew Cappucci and Andrew Freedman, "Europe to See Third Major Heat Wave This Summer, as Temperatures Soar from France to Scandinavia," *Washington Post*, August 22, 2019, https://www.washingtonpost.com/weather/2019/08/22/europe-see-third-major-heat-wave-this-year-temperatures-soar-france-scandinavia/.

4 *"500-year storms" in the last five years*: Amal Ahmed, "Tropical Storm Imelda Will Likely Be Southeast Texas' Fifth 500-Year Flood in Five Years," *Texas Observer*, September 20, 2019, https://www.texasobserver.org/tropical-storm-imelda-will-likely-be-southeast-texas-fifth-500-year-flood-in-five-years/.

5 *an eye-opening report*: V. Masson-Delmotte et al., eds., "Summary for Policy-makers," in *Global Warming of 1.5°C: An IPCC Special Report on the Impacts of Global Warming of 1.5°C Above Pre-industrial Levels and Related Global Greenhouse Gas Emission Pathways, in the Context of Strengthening the Global Response to the Threat of Climate Change, Sustainable Development, and Efforts to Eradicate Poverty*," IPCC, 2018, https://www.ipcc.ch/sr15/chapter/spm/.

6 *the largest historical share of carbon emissions*: Hannah Ritchie, "Who has contributed most to global CO_2 emissions?" Our World in Data, October 1, 2019, https://ourworldindata.org/contributed-most-global-co2.

6 *a "fair share" of US action would have us decarbonizing over the next decade*: William Lynn, "Who's Most Responsible for Climate Change?," *New Republic*, December 8, 2015, https://newrepublic.com/article/125279/whos-responsible-climate-change.

6 *in a world warmed by 1.5 degrees*: Drew Shindell, Greg Faluvegi, Karl Seltzer, and Cary Shindell, "Quantified, Localized Health Benefits of Accelerated Carbon Dioxide Emissions Reductions," *Nature*, March 19, 2018, https://www.nature.com/articles/s41558-018-0108-y.

6 *the UN expects*: H.-O. Pörtner et al., eds., "Summary for Policymakers," in *IPCC Special Report on the Ocean and Cryosphere in a Changing Climate*, IPCC, 2019, https://www.ipcc.ch/srocc/.

6 *will be unlivable in summer*: Eun-Soon Im, Jeremy S. Pal, and Elfatih A. B. Eltahir, "Deadly Heat Waves Projected in the Densely Populated Agricultural Regions of South Asia," *Science Advances*, August 2, 2017, https://advances.sciencemag.org/content/3/8/e1603322.full; Jeremy S. Pal and Elfatih A. B. Eltahir, "Future Temperature in Southwest Asia Projected to Exceed a Threshold for Human Adaptability," *Nature*, October 26, 2015, https://www.nature.com/articles/nclimate2833.

7 *200 million or more climate refugees*: Oli Brown, *Migration and Climate Change* 31 (January 2008), IOM Migration Research Series, International Organization for Migration, 2008, https://www.ipcc.ch/apps/njlite/srex/njlite_download.php?id=5866, 9, 11, 28–29, 32.

7 *280 million people*: Pörtner, "Summary for Policymakers."

7 *under what conditions that new life will burn*: David Wallace-Wells, "Los Angeles Fire Season Is Beginning Again. And It Will Never End. A Bulletin from Our Climate Future," *New York Magazine*, May 12, 2019, https://nymag.com/intelligencer/2019/05/los-angeles-fire-season-will-never-end.html.

7 *loss of all the planet's ice sheets*: Fred Pearce, "As Climate Change Worsens, a Cascade of Tipping Points Looms," *Yale Environment 360*, December 5, 2019, https://e360.yale.edu/features/as-climate-changes-worsens-a-cascade-of-tipping-points-looms.

7 *more than 200 feet*: "Quick Facts on Ice Sheets," National Snow and Ice Data Center, 2020, https://nsidc.org/cryosphere/quickfacts/icesheets.html.

8 *an impact twice as deep as the Great Depression, and permanent*: Marshall Burke, W. Matthew Davis, and Noah S. Diffenbaugh, "Large Potential Reduction in Economic Damages Under UN Mitigation Targets," Research Letter, *Nature*, May 2018, 549–53, https://web.stanford.edu/~mburke/papers/BurkeDavisDiffenbaugh2018.pdf.

8 *deep as the Great Depression, and permanent*: Felix Salmon, "The Cost of Climate Change," *Axios*, October 14, 2018, https://www.axios.com/climate-change-costs-wealth-carbon-tax-303d7cff-3085-49d9-accb-ec77689b9911.html.

8 *producing widespread famine*: Christopher Flavelle, "Climate Change Threatens the World's Food Supply, United Nations Warns," *New York Times*, August 8, 2019, https://www.nytimes.com/2019/08/08/climate/climate-change-food-supply.html.

8 *direct heat, desertification, and flooding*: Gaia Vince, "How to Survive the Coming Century," *New Scientist*, February 25, 2009, https://www.newscientist.com/article/mg20126971-700-how-to-survive-the-coming-century/.

8 *more than twice as much war*: Solomon M. Hsiang, Marshall Burke, and Edward Miguel, "Quantifying the Influence of Climate on Human Conflict," *Science*, September 13, 2013, https://science.sciencemag.org/content/341/6151/1235367.

8 *and half as much food*: David S. Battisti and Rosamond L. Naylor, "Historical Warnings of Future Food Insecurity with Unprecedented Seasonal Heat," *Science*, January 9, 2009, https://science.sciencemag.org/content/323/5911/240.full.

9 *in just the past three decades*: T. A. Boden, G. Marland, and R. J. Andres, "Global, Regional, and National Fossil-Fuel CO_2 Emissions," Carbon Dioxide Information Analysis Center, 2017, https://cdiac.ess-dive.lbl.gov/trends/emis/overview_2014.html.

11 *We are now burning 60 percent more coal than we were just in the year 2000*: "Coal Information 2019," International Energy Agency, August 2019, accessed March 2020, https://www.iea.org/reports/coal-information-2019.

CHAPTER 2: *WE DIDN'T START THE FIRE* / KATE ARONOFF

14 *former British prime minister Margaret Thatcher said*: Margaret Thatcher, interviewed by Douglas Keay, *Woman's Own*, September 23, 1987, https://www.margaretthatcher.org/document/106689.

15 *"I'm from the government and I'm here to help"*: Ronald Reagan, news conference, August 12, 1986, https://www.reaganfoundation.org/ronald-reagan/reagan-quotes-speeches/news-conference-1/.

15 *"Call me a pessimist"*: Jonathan Franzen, "What If We Stopped Pretending?," *The New Yorker*, September 8, 2019, https://www.newyorker.com/culture/cultural-comment/what-if-we-stopped-pretending.

15 *"We have trained ourselves"*: Nathaniel Rich, "Losing Earth: The Decade We Almost Stopped Climate Change," *New York Times Magazine*, August 1, 2018, https://www.nytimes.com/interactive/2018/08/01/magazine/climate-change-losing-earth.html?mtrref=www.google.com&assetType=REGIWALL#main.

16 *the world's sixth-largest polluter*: Matthew Taylor and Jonathan Watts, "Revealed: The 20 Firms Behind a Third of All Carbon Emissions," *The Guardian*, October 9, 2019, https://www.theguardian.com/environment/2019/oct/09/revealed-20-firms-third-carbon-emissions.

17 *pay a lower tax rate than the bottom 50 percent of households*: Christopher Ingraham, "Wealth Concentration Returning to 'Levels Last Seen During the Roaring Twenties,' According to New Research," *Washington Post*, February 8, 2019, https://www.washingtonpost.com/us-policy/2019/02/08/wealth-concentration-returning-levels-last-seen-during-roaring-twenties-according-new-research/#targetText=Wealth%2C%20here%2C%20is%20roughly%20synonymous,the%20value%20of%20any%20debt.

17 *the wealthiest 10 percent of the world's population*: "Extreme Carbon Inequality: Why the Paris Climate Deal Must Put the Poorest, Lowest Emitting and Most Vulnerable People First," Oxfam, December 2, 2015, https://oi-files-d8-prod.s3.eu-west-2.amazonaws.com/s3fs-public/file_attachments/mb-extreme-carbon-inequality-021215-en.pdf.

17 *just ninety corporations*: Richard Heede, "Tracing Anthropogenic Carbon Dioxide and Methane Emissions to Fossil Fuel and Cement Producers, 1854–2010," *Climatic Change* 122 (2013): 229–41, https://link.springer.com/article/10.1007/s10584-013-0986-y.

17 *35 percent of the world's energy-related carbon dioxide and methane emissions*: Taylor and Watts, "Revealed: The 20 Firms Behind a Third of All Carbon Emissions."

17 *researcher Dario Kenner found*: Dario Kenner, *Carbon Inequality: The Role of the Richest in Climate Change* (Abingdon, UK: Routledge, 2019); Kate Aronoff, "Jay Inslee just dropped the most ambitious climate plan from a presidential candidate. Here's who it targets," *The Intercept*, June 24, 2019, https://theintercept.com/2019/06/24/jay-inslee-climate-change-pollution/; Dario Kenner, "The Polluter Elite Database," June 2019, https://whygreeneconomy.org/the-polluter-elite-database/.

17 *as executives like Tillerson do*: Jad Mouawad, "The New Face of an Oil Giant," *New York Times*, March 30, 2006, https://www.nytimes.com/2006/03/30/business/the-new-face-of-an-oil-giant.html.

17 *ExxonMobil and Shell misled the public*: Benjamin Franta, "Shell and Exxon's Secret 1980s Climate Change Warnings," *The Guardian*, September 19, 2008, https://www.theguardian.com/environment/climate-consensus-97-per-cent/2018/sep/19/shell-and-exxons-secret-1980s-climate-change-warnings.

17 *climate denier groups like the Heartland Institute*: "Smoke, Mirrors, and Hot Air," Union of Concerned Scientists, January 2007, https://web.archive.org/web/20150726204316/http://www.ucsusa.org/sites/default/files/legacy/assets/documents/global_warming/exxon_report.pdf; Naomi Oreskes and Erik M. Conway, *Merchants of Doubt: How a Handful of Scientists*

Obscured the Truth on Issues from Tobacco Smoke to Global Warming (New York: Bloomsbury Publishing, 2010), 169–274.

18 *gin up a debate in the mainstream media*: Oreskes and Conway, *Merchants of Doubt*, 169–274.

18 *transform the Republican Party*: Suzanne Goldenberg, "ExxonMobil Gave Millions to Climate-Denying Lawmakers Despite Pledge," *The Guardian*, July 15, 2015, https://www.theguardian.com/environment/2015/jul/15/exxon-mobil-gave-millions-climate-denying-lawmakers.

18 *lobby against US involvement in the Kyoto Protocol*: Janet Sawin and Kert Davies, "Denial and Deception: A Chronicle of ExxonMobil's Efforts to Corrupt the Debate on Global Warming," Greenpeace, May 2002, https://www.greenpeace.org/usa/wp-content/uploads/2015/11/exxon-denial-and-deception.pdf?a1481f; Climate Investigations Center, "The Global Climate Coalition: Big Business Funds Climate Change Denial and Regulatory Delay," March 25, 2019, https://climateinvestigations.org/wp-content/uploads/2019/04/The-Global-Climate-Coalition-Denial-and-Delay.pdf.

18 *estimated $5.1 trillion in direct and indirect subsidies worldwide*: David Coady, Ian Parry, Nghia-Piotr Le, and Baoping Shang, "Global Fossil Fuel Subsidies Remain Large: An Update Based on Country-Level Estimates," International Monetary Fund, May 2, 2019, https://www.imf.org/en/Publications/WP/Issues/2019/05/02/Global-Fossil-Fuel-Subsidies-Remain-Large-An-Update-Based-on-Country-Level-Estimates-46509.

18 *"rapid and far-reaching transitions"*: V. Masson-Delmotte et al., eds., "Summary for Policymakers," in *Global Warming of 1.5°C: An IPCC Special Report on the Impacts of Global Warming of 1.5°C Above Pre-industrial Levels and Related Global Greenhouse Gas Emission Pathways, in the Context of Strengthening the Global Response to the Threat of Climate Change, Sustainable Development, and Efforts to Eradicate Poverty*," IPCC, 2018, https://www.ipcc.ch/sr15/chapter/spm/.

19 *global emissions reductions of 7.6 percent*: "Emissions Gap Report 2019," United Nations Environment Programme, 2019, https://www.unenvironment.org/resources/emissions-gap-report-2019.

19 *collapse of the Soviet Union*: Quirin Schiermeier, "Soviet Union's Collapse Led to Massive Drop in Carbon Emissions," *Nature*, July 1, 2019, https://www.nature.com/articles/d41586-019-02024-6; Matthew L. Wald, "Carbon Dioxide Emissions Dropped in 1990, Ecologists Say," *New York Times*, December 8, 1991, https://www.nytimes.com/1991/12/08/world/carbon-dioxide-emissions-dropped-in-1990-ecologists-say.html.

19 *decline by 97, 87, and 74 percent*: Emma Foehringer Merchant, "IPCC: Renewables to Supply 70%–85% of Electricity by 2050 to Avoid Worst Impacts of Climate Change," Greentech Media, October 8, 2018, https://www.greentechmedia.com/articles/read/ipcc-renewables-85-electricity-worst-impacts-climate-change#gs.3g1obl.

19 *stubborn, well-documented link*: Carolyn Korman, "The False Choice Between Economic Growth and Combatting Climate Change," *The New Yorker*, February 4, 2019, https://www.newyorker.com/news/news-desk/the-false-choice-between-economic-growth-and-combatting-climate-change.

20 *economic activity that takes place on the electricity grid*: "Powering America: The Economic and Workforce Contributions of the US Electric Power Industry," M. J. Bradley & Associates, 2017, https://mjbradley.com/about-us/case-studies/powering-america.

20 *plant billions or even trillions of trees*: Jean-Francois Bastin et al., "The Global Tree Restoration Potential," *Science*, July 5, 2019, https://doi.org/10.1126/science.aax0848.

21 *like those in place at the state level in California*: "California's Renewables Portfolio Standard (RPS) Program," Union of Concerned Scientists, July 2016, https://www.ucsusa.org/resources/californias-renewables-portfolio-standard-program#ucs-report-downloads.

22 *has remained largely flat*: "Short-Term Energy Outlook (STEO)," US Energy Information Administration, March 2020, https://www.eia.gov/outlooks/steo/report/electricity.php.

22 *fossil fuel extraction has continued to increase*: Hannah Ritchie and Max Roser, "Fossil Fuels," Our World in Data, 2020, https://ourworldindata.org/fossil-fuels.

22 *United States becomes a net exporter*: Bradley Olson, "US Becomes Net Exporter of Oil, Fuels for First Time in Decades," *Wall Street Journal*, December 6, 2018, https://www.wsj.com/articles/u-s-becomes-net-exporter-of-oil-fuels-for-first-time-in-decades-1544128404.

22 *roughly $20 billion in subsidies*: Dana Nuccatelli, "America Spends over $20bn Per Year on Fossil Fuel Subsidies. Abolish Them," *The Guardian*, July 30, 2018, https://www.theguardian.com/environment/climate-consensus-97-per-cent/2018/jul/30/america-spends-over-20bn-per-year-on-fossil-fuel-subsidies-abolish-them.

22 *Permanent tax breaks to the fossil fuel sector*: David Roberts, "Friendly Policies Keep US Oil and Coal Afloat Far More Than We Thought," *Vox*, July 26,

2018, https://www.vox.com/energy-and-environment/2017/10/6/16428458/us-energy-coal-oil-subsidies.

22 *example already set by New Zealand*: David Reid, "New Zealand Set to Ban New Offshore Oil and Gas Drilling," CNBC, April 12, 2018, https://www.cnbc.com/2018/04/12/new-zealand-set-to-ban-oil-and-gas-drilling.html.

22 *crude oil export ban, lifted in 2015*: "Oil's Well That Ends Well: America Lifts Its Ban on Oil Exports," *The Economist*, December 18, 2015, https://www.economist.com/finance-and-economics/2015/12/18/america-lifts-its-ban-on-oil-exports.

22 *as much as half of new oil and gas development*: Roberts, "Friendly Policies Keep US Oil and Coal Afloat Far More Than We Thought."

23 *researchers at the Next System Project*: Carla Skandier, "Quantitative Easing for the Planet," The Next System Project, August 30, 2018, https://thenextsystem.org/learn/stories/quantitative-easing-planet.

23 *ensuring a dignified quality of life*: Jeremy Brecher, "Making the Green New Deal Work for Workers," *In These Times*, April 22, 2019, https://inthesetimes.com/features/green-new-deal-worker-transition-jobs-plan.html.

23 *start at $40 per ton*: "Report of the High-Level Commission on Carbon Prices," Carbon Pricing Leadership Coalition, May 29, 2017, https://www.carbonpricingleadership.org/report-of-the-highlevel-commission-on-carbon-prices.

24 global *carbon tax ranging from $135 to $5,500 per ton*: Masson-Delmotte et al., "Summary for Policymakers."

24 *highest carbon tax on earth*: "Sweden's Carbon Tax," Government Offices of Sweden, February 2020, https://www.government.se/government-policy/taxes-and-tariffs/swedens-carbon-tax/.

24 *existing prices on carbon worldwide*: Brad Plumer, "New U.N. Climate Report Says Put a High Price on Carbon," *New York Times*, October 8, 2018, https://www.nytimes.com/2018/10/08/climate/carbon-tax-united-nations-report-nordhaus.html.

24 *France attempted to impose*: Adam Nossiter, "France Suspends Fuel Tax Increase That Spurred Violent Protests," *New York Times*, December 4, 2018, https://www.nytimes.com/2018/12/04/world/europe/france-fuel-tax-yellow-vests.html.

25 *The economist Rexford Tugwell, a member of FDR's so-called brain trust*: Rexford G. Tugwell, "Design for Government," *Political Science Quarterly* 48, no. 3 (1933): 321–32, https://doi.org/10.2307/2143150.

CHAPTER 3: *MARKET FUNDAMENTALISM AT THE WORST TIME* / NAOMI KLEIN

28 *becoming cheaper, more efficient*: Damian Carrington, "Solar Power Drives Renewable Energy Investment Boom in 2014," *The Guardian*, January 9, 2015, https://www.theguardian.com/environment/2015/jan/09/solar-power-drives-renewable-energy-investment-boom-2014.

28 *growing victory gardens in 1943*: Sarah Sundin, "Victory Gardens in World War II," University of California Master Gardener Program of Sonoma County, http://sonomamg.ucanr.edu/History/Victory_Gardens_in_World_War_II/.

30 *talking seriously about radical cuts to greenhouse gas emissions in 1988*: Spencer Weart, "A Hyperlinked History of Climate Change Science," *The Discovery of Global Warming*, July 2017, https://history.aip.org/climate/summary.htm.

31 *rapid growth rate continues to this day*: Corinne Le Quéré, Michael R. Raupach, and Ian Woodward, "Trends in the Sources and Sinks of Carbon Dioxide," *Nature Geoscience* 2 (December 2009): 831–36, https://www.nature.com/articles/ngeo689#citeas.

31 *saw the largest absolute increase since the Industrial Revolution*: Glen P. Peters et al., "Rapid Growth in CO_2 Emissions After the 2008–2009 Global Financial Crisis," *Nature Climate Change* 2 (2012), https://www.nature.com/articles/nclimate1332.

32 *mass export of products*: "Exports of Goods and Services (Current US$)," World Bank, 2019, https://data.worldbank.org/indicator/NE.EXP.GNFS.CD?end=2018&start=1960&view=chart.

32 *a uniquely wasteful model of production, consumption, and agriculture*: Mark Notaras, "Agriculture and Food Systems Unsustainable," United Nations University, June 30, 2010, https://ourworld.unu.edu/en/agriculture-and-food-systems-unsustainable.

32 *effects are cumulative, growing more severe with time*: Alice Bows and Kevin Anderson, "Beyond 'Dangerous' Climate Change: Emission Scenarios for a New World," *Philosophical Transactions of the Royal Society A* 369 (2011), 20–44, https://royalsocietypublishing.org/doi/full/10.1098/rsta.2010.0290.

32 *cut their emissions by somewhere in the neighborhood of 8 to 10 percent a year*: Ibid.; Kevin Anderson, "EU 2030 Decarbonisation Targets and UK Carbon Budgets: Why So Little Science?," Kevinanderson.info, June 14, 2013, http://kevinanderson.info/blog/eu-2030-decarbonisation-targets-and-uk-carbon-budgets-why-so-little-science/.

33 *authored a landmark report*: Gro Harlem Brundtland et al., "Environment

and Development Challenges: The Imperative to Act," Conservation International, https://www.conservation.org/docs/default-source/publication-pdfs/ci_rioplus20_blue-planet-prize_environment-and-development-challenges.pdf.

34 *"Our ongoing and collective carbon profligacy"*: Kevin Anderson, "Why Carbon Prices Can't Deliver the 2°C Target," Kevinanderson.info, August 13, 2013, http://kevinanderson.info.

34 *as US president Barack Obama described his approach*: Jason Furman and Jim Stock, "New Report: The All-of-the-Above Energy Strategy as a Path to Sustainable Economic Growth," The White House, May 29, 2014, https://obamawhitehouse.archives.gov/blog/2014/05/29/new-report-all-above-energy-strategy-path-sustainable-economic-growth.

34 *"we have no room to build anything that emits CO_2 emissions"*: Quoted in Adam Vaughan, "World Has No Capacity to Absorb New Fossil Fuel Plants, Warns IEA," *The Guardian*, November 12, 2019, https://www.theguardian.com/business/2018/nov/13/world-has-no-capacity-to-absorb-new-fossil-fuel-plants-warns-iea.

35 *eaten up by existing fossil fuel infrastructure*: Ibid.

35 *he edited a special issue*: Gary Stix, "A Climate Repair Manual," *Scientific American*, September 1, 2006, https://www.scientificamerican.com/article/a-climate-repair-manual/.

35 *in 2012 Stix wrote*: Gary Stix, "Effective World Government Will Be Needed to Stave Off Climate Catastrophe," *Scientific American*, March 17, 2012, https://blogs.scientificamerican.com/observations/effective-world-government-will-still-be-needed-to-stave-off-climate-catastrophe/.

CHAPTER 4: *AVERTING CLIMATE COLLAPSE REQUIRES CONFRONTING RACISM* / IAN HANEY LÓPEZ

38 *racism at the scale that science and justice demand*: "Green New Deal," Sunrise Movement, https://www.sunrisemovement.org/green-new-deal.

40 *four out of five, 84 percent, denied that climate change is caused by humans*: Heather Smith, "Climate Deniers Are More Likely to Be Racist. Why?," *Sierra*, June 18, 2018, Salil D. Benegal, "The Spillover of Race and Racial Attitudes into Public Opinion About Climate Change," *Environmental Politics* 27, no. 4 (published online March 27, 2018).

40 *blocking climate action that would reduce the profits of the world's largest polluters*: Coral Davenport, "Climate Change Denialists in Charge," *New York Times*, March 27, 2017.

41 *dog whistle politics played an outsize role in destroying the original New Deal*: Ian Haney López, *Dog Whistle Politics: How Coded Racial Appeals Have Reinvented Racism and Wrecked the Middle Class* (New York: Oxford University Press, 2014).

43 *his race-baiting as "a blueprint for everything I've done in the South"*: Tali Mendelberg, *The Race Card: Campaign Strategy, Implicit Messages, and the Norm of Equality* (Princeton, NJ: Princeton University Press, 2001), 143.

43 *"You start out in 1954 by saying, 'Nigger, nigger, nigger . . .'"*: Rick Perlstein, "Exclusive: Lee Atwater's Infamous 1981 Interview on the Southern Strategy," TheNation.com, November 13, 2012.

43 *From "nigger, nigger, nigger" to "states' rights" and "forced busing"*: What follows in the essay draws on Ian Haney López, *Merge Left: Fusing Race and Class, Winning Elections, and Saving America* (New York: The New Press, 2019).

44 *1. Fear and resent people of color; 2. Distrust government; 3. Trust the marketplace*: Ian Haney López, *Merge Left: Fusing Race and Class, Winning Elections, and Saving America*.

47 *It was devastatingly effective from behind the scenes*: Christopher Leonard, *Kochland: The Secret History of Koch Industries and Corporate Power in America* (New York: Simon & Schuster, 2019).

47 *anything from happening back when there was still time*: Jane Mayer, "'Kochland' Examines the Koch Brothers' Early, Crucial Role in Climate-Change Denial," *The New Yorker*, August 13, 2019.

47 *circle of wealthy donors they enlisted to fight for rule by the rich*: Jane Mayer, *Dark Money: The Hidden History of the Billionaires Behind the Rise of the Radical Right*, repr. ed. (New York: Anchor, 2016).

48 *"And it's a revolt against our black President"*: Matthew Rothschild, "Rampant Xenophobia," *The Progressive*, October 2010.

48 "*and the frogs come out of the mud—and they're our candidates!*": Quoted in Jane Mayer, "Covert Operations: The Billionaire Brothers Who Are Waging a War Against Obama," *The New Yorker*, August 23, 2010.

48 "*helped turn their private agenda into a mass movement*": Ibid.

48 *"grassroots citizens' movement brought to you by a bunch of oil billionaires"*: Ibid.

48 *stop climate legislation, their main tactic was to fund racial division*: Jane Mayer, "Koch Pledge Tied to Congressional Climate Inaction," *The New Yorker*, June 30, 2013.

49 "*It's not your government anymore; it's theirs*": Arlie Russell Hochschild, "I

Spent 5 Years with Some of Trump's Biggest Fans. Here's What They Won't Tell You," *Mother Jones*, September/October 2016.

49 *her narrative records the essential teachings of dog whistle politics*: See generally Arlie Hochschild, *Strangers in Their Own Land: Anger and Mourning on the American Right* (New York: The New Press, 2016), chapter 9.

49 "*Trump has delivered for the Kochs*": Mayer, " 'Kochland' Examines the Koch Brothers' Early, Crucial Role in Climate-Change Denial."

49 *"more progress in the last five years than I had in the previous fifty"*: Philip Elliott, "The Koch Brothers Plan to Spend a Record-Setting $400 Million," *Time*, January 27, 2018.

50 "*more affordable for people struggling to make ends meet*": Ian Haney López, "Race-Class Narrative National Dial Survey Report," May 2018, https://www.ianhaneylopez.com/race-class-project/.

CHAPTER 5: *HOW WE GOT TO THE GREEN NEW DEAL* / BILL MCKIBBEN

55 *historic testimony before Congress*: James Shabecoff, "Global Warming Has Begun, Expert Tells Senate," *New York Times*, June 24, 1988, https://www.nytimes.com/1988/06/24/us/global-warming-has-begun-expert-tells-senate.html.

55 *first book for a general audience on the topic*: Bill McKibben, *The End of Nature* (New York: Random House, 1989).

56 *"with the White House effect"*: Cass Peterson, "EXPERTS, OMB SPAR ON GLOBAL WARMING," *Washington Post*, May 9, 1989.

56 *called for a tax on the greenhouse gas emissions*: Bernard P. Herber and Jose T. Raga, "An International Carbon Tax to Combat Global Warming: An Economic and Political Analysis of the European Union Proposal," *American Journal of Economics and Sociology* 54, no. 3 (1995): 257–67, www.jstor.org/stable/3487089.

56 *than in all of human history before that time*: Richard Heede, "Tracing Anthropogenic Carbon Dioxide and Methane Emissions to Fossil Fuel and Cement Producers, 1854–2010," *Climatic Change* 122 (2013): 229–41, https://link.springer.com/article/10.1007/s10584-013-0986-y.

57 InsideClimate News: Neela Banerjee, Lisa Song, and David Hasemyer, "Exxon: The Road Not Taken," *InsideClimate News*, September 26, 2015, https://insideclimatenews.org/content/Exxon-The-Road-Not-Taken.

57 Los Angeles Times: Amy Lieberman and Susanne Rust, "Big Oil Braced for Global Warming While It Fought Regulations," *Los Angeles Times*, December 31, 2015, https://graphics.latimes.com/oil-operations/#about.

57 *Columbia Journalism School*: Sara Jerving, Katie Jennings, Masako Melissa Hirsch, Susanne Rust, Dino Gandoni, Amy Lieberman, Asaf Shalev, Michael Phillis, and Elah Feder, "Two-Year Long Investigation: What Exxon Knew About Climate Change," Columbia Journalism School Energy and Environmental Reporting Project, 2017, https://journalism.columbia.edu/two-year-long-investigation-what-exxon-knew-about-climate-change.

57 *Exxon began building its drilling rigs higher*: Amy Lieberman and Susanne Rust, "Big Oil Braced for Global Warming While It Fought Regulations."

57 *pump out disinformation*: Janet Sawin, Kert Davies, Greenpeace United Kingdom, Ross Gelbspan, Kirsty Hamilton, and Bill Hare, "Denial and Deception: A Chronicle of ExxonMobil's Efforts to Corrupt the Debate on Global Warming," Greenpeace, May 2002, https://www.greenpeace.org/usa/wp-content/uploads/2015/11/exxon-denial-and-deception.pdf?a1481f.

57 *many veterans of the tobacco wars:* Benjamin Hulac, "Tobacco and Oil Industries Used Same Researchers to Sway Public," *Scientific American*, July 20, 2016, https://www.scientificamerican.com/article/tobacco-and-oil-industries-used-same-researchers-to-sway-public1/.

58 *the planet was cooling*: Lee R. Raymond, "Energy—Key to Growth and a Better Environment for Asia-Pacific Nations," Exxon Corporation, October 13, 1997, http://www.climatefiles.com/exxonmobil/1997-exxon-lee-raymond-speech-at-world-petroleum-congress/.

58 *set records for new profits*: Matthew Taylor and Jillian Ambrose, "Revealed: Big Oil's Profits Since 1990 Total Nearly $2tn," *The Guardian*, February 12, 2020, https://www.theguardian.com/business/2020/feb/12/revealed-big-oil-profits-since-1990-total-nearly-2tn-bp-shell-chevron-exxon.

58 *It was just days into Bush's presidency*: Steve Coll, *Private Empire: ExxonMobil and American Power* (New York: Penguin Books, 2012).

59 *the decade of Al Gore's* An Inconvenient Truth: Al Gore, *An Inconvenient Truth: The Crisis of Global Warming*, rev. ed. (New York: Viking, 2007).

59 *essentially nothing happened*: John Vidal, Allegra Stratton, and Suzanne Goldenberg, "Low Targets, Goals Dropped: Copenhagen Ends in Failure," *The Guardian*, December 18, 2009, https://www.theguardian.com/environment/2009/dec/18/copenhagen-deal.

59 *Senate didn't even bother to hold a vote*: Brian C. Black, "Waxman-Markey Climate Bill: American Clean Energy and Security Act of 2009," in *Climate Change: An Encyclopedia of Science and History*, ed. Brian C. Black et al., Vol. 4: 1405–1407 (Santa Barbara, CA: ABC-CLIO, 2013), https://

link.gale.com/apps/doc/CX2721900236/GVRL?u=wash_main&sid =GVRL&xid=260fcec4.

59 *reject TransCanada's construction permit*: Coral Davenport, "Citing Climate Change, Obama Rejects Construction of Keystone XL Oil Pipeline," *New York Times*, November 6, 2015, https://www.nytimes.com/2015/11/07/us /obama-expected-to-reject-construction-of-keystone-xl-oil-pipeline.html.

59 *insisted that water is life*: Julia C. Wong, "Dakota Access Pipeline: 300 Protesters Injured After Police Use Water Cannons," *The Guardian*, November 21, 2016, https://www.theguardian.com/us-news/2016/nov/21/dakota -access-pipeline-water-cannon-police-standing-rock-protest.

59 *credits the Dakota Access fight*: Rebecca Solnit, "Standing Rock Inspired Ocasio-Cortez to Run. That's the Power of Protest," *The Guardian*, January 14, 2019, https://www.theguardian.com/commentisfree/2019/jan/14 /standing-rock-ocasio-cortez-protest-climate-activism.

60 *meant that the wealth of the 1 percenters*: Bradford A. Lee, "The New Deal Reconsidered," *Wilson Quarterly* 6, no. 2 (1982): 62–76, https://www.jstor .org/stable/40256265; Emmanuel Saez and Gabriel Zucman, "Wealth Inequality in the United States Since 1913: Evidence from Capitalized Income Tax Data," Quarterly Journal of Economics 131, no. 2 (2016): 519–78, https://doi.org/10.1093/qje/qjw004.

60 *the lowest it has ever been*: Emmanuel Saez and Gabriel Zucman, "Wealth Inequality in the United States Since 1913: Evidence from Capitalized Income Tax Data," *Quarterly Journal of Economics* 131, no. 2 (2016): 519–78, https://doi.org/10.1093/qje/qjw004.

60 *head of the Federal Reserve*: Christopher Hitchens, "Greenspan Shrugged," *Vanity Fair*, December 6, 2000, https://www.vanityfair.com/culture/2000 /12/hitchens-200012.

60 *bottom 3.6 billion people*: Deborah Hardoon, "An Economy for the 99%," Oxfam, January 16, 2017, https://www.oxfam.org/en/research/economy-99.

60 *more money than the bottom 40 percent of the American population*: Sylvia Allegretto, "One Step Up and Two Steps Back," *Berkeley Blog*, UC Berkeley, October 2, 2014, https://blogs.berkeley.edu/2014/10/02/one-step-up-two -steps-back/.

60 *to get them through an emergency*: Cameron Huddleston, "Survey: 69% of Americans Have Less Than $1,000 in Savings," GOBankingRates, December 16, 2019, https://www.gobankingrates.com/saving-money/savings-advice /americans-have-less-than-1000-in-savings/.

61 *blocking renewable energy*: Evan Halper, "Koch Brothers, Big Utilities Attack

Solar, Green Energy Policies," *Los Angeles Times*, April 19, 2014, https://www.latimes.com/nation/la-na-solar-kochs-20140420-story.html.

61 *defunding mass transit*: Hiroko Tabuchi, "How the Koch Brothers Are Killing Public Transit Projects Around the Country," *New York Times*, June 18, 2018, https://www.nytimes.com/2018/06/19/climate/koch-brothers-public-transit.html.

61 *trashing environmentalists*: Suzanne Goldenberg and Ed Pilkington, "ALEC Calls for Penalties on 'Freerider' Homeowners in Assault on Clean Energy," *The Guardian*, December 4, 2013, https://www.theguardian.com/world/2013/dec/04/alec-freerider-homeowners-assault-clean-energy.

62 *taxes were an insupportable burden*: Adam Nossiter, "France Suspends Fuel Tax Increase That Spurred Violent Protests," *New York Times*, December 4, 2018, https://www.nytimes.com/2018/12/04/world/europe/france-fuel-tax-yellow-vests.html.

62 *carbon price in the state of Washington*: Marianne Lavelle, "Big Oil Has Spent Millions of Dollars to Stop a Carbon Fee in Washington State," *InsideClimate News*, October 29, 2018, https://insideclimatenews.org/news/29102018/election-2018-washington-carbon-fee-ballot-initiative-price-carbon-big-oil-opposition.

62 *people's backyards and school zones*: Umair Irfan, "A Major Anti-Fracking Ballot Measure in Colorado Has Failed," *Vox*, November 7, 2018, https://www.vox.com/2018/11/5/18064604/colorado-election-results-fracking-proposition-112.

65 *"time makes ancient good uncouth"*: James R. Lowell, "Once to Every Man and Nation," *Boston Courier*, December 11, 1845, https://www.greatchristianhymns.com/once-every-man.html.

65 *dying of the Great Barrier Reef*: T. P. Hughes, J. T. Kerry, and T. Simpson, "Large-Scale Bleaching of Corals on the Great Barrier Reef," *Ecology* 99, no. 2 (2017), 10.1002/ecy.2092.

65 *the burning of the Amazon*: Herton Escobar, "Amazon Fires Clearly Linked to Deforestation, Scientists Say," *Science,* August 30, 2019, https://science.sciencemag.org/content/365/6456/853.

65 *400 parts per million and then 410*: "Monthly Average Mauna Loa CO_2," Global Monitoring Division, Earth System Research Laboratory, March 5, 2020, https://www.esrl.noaa.gov/gmd/ccgg/trends/.

65 *biggest rainfalls in American history*: "Assessing the US Climate in June 2019," National Oceanic and Atmospheric Administration, July 9, 2019, https://www.ncei.noaa.gov/news/national-climate-201906.

65 *turned into hell inside half an hour*: E. A. Williams, "At Least Nine Dead, Paradise 'Pretty Much Destroyed' as Wildfire Rages in Northern California," *Washington Post*, November 9, 2018, https://www.washingtonpost.com/weather/2018/11/09/town-called-paradise-pretty-much-destroyed-wildfire-rages-northern-california/?itid=lk_inline_manual_2.

65 *diseases of despair take their toll*: Joshua Cohen, "'Diseases of Despair' Contribute to Declining US Life Expectancy," *Forbes*, July 19, 2018, https://www.forbes.com/sites/joshuacohen/2018/07/19/diseases-of-despair-contribute-to-declining-u-s-life-expectancy/#7bdf06dd656b.

65 "*but it bends toward justice*": Martin Luther King Jr., "Statement on Ending the Bus Boycott," December 20, 1956, Montgomery, Alabama, https://kinginstitute.stanford.edu/king-papers/documents/statement-ending-bus-boycott#fn1.

CHAPTER 6: *POLICIES AND PRINCIPLES OF A GREEN NEW DEAL* / RHIANA GUNN-WRIGHT

69 *"Policy [is] a statement by government . . ."*: Thomas A. Birkland, *An Introduction to the Policy Process: Theories, Concepts, and Models of Public Policy Making*, 4th ed. (New York: Routledge, 2005), 8, 9.

73 *low-income workers, women, the elderly, the unhoused*: Recognizing the Duty of the Government to Create a Green New Deal, H.R. 109, February 7, 2019, Section 1, https://www.congress.gov/116/bills/hres109/BILLS-116hres109ih.pdf.

74 *especially from high-emitting countries like the US*: "Global Warming of 1.5°C," Intergovernmental Panel on Climate Change, 2018, https://report.ipcc.ch/sr15/pdf/sr15_spm_final.pdf; David Wallace-Wells, "What If the Courts Could Save the Climate?," *New York Magazine*, November 29, 2018, https://nymag.com/intelligencer/2018/11/julianna-v-united-states-how-courts-could-save-the-climate.html.

76 *discouraged lending and encouraged redlining*: Bruce Mitchell and Juan Franco, "HOLC 'Redlining' Maps: The Persistent Structure of Segregation and Economic Inequality," National Community Reinvestment Coalition, March 20, 2018, https://ncrc.org/wp-content/uploads/dlm_uploads/2018/02/NCRC-Research-HOLC-10.pdf.

76 *erasing decades of wealth for those who owned homes and businesses*: "Renewing Inequality," Digital Scholarship Lab, American Panorama, https://dsl.richmond.edu/panorama/renewal/#view=0/0/1&viz=cartogram&city=chicagoIL&loc=11/41.8640/-87.6340.

77 *the assumptions of neoliberalism*: David Harvey, *A Brief History of Neoliberalism* (New York: Oxford University Press, 2017); Kim Phillips-Fein, *Invisible Hands: The Businessmen's Crusade Against the New Deal* (New York: W. W. Norton & Company, 2010).

78 *"actively creating and shaping (new) markets, while regulating existing ones"*: Mariana Mazzucato, *The Entrepreneurial State: Debunking Public vs. Private Sector Myths* (London: Anthem Press, 2013), 5, 6.

79 *public spending for R&D declined nearly 50 percent since the 1980s*: J. John Wu, "Why US Business R&D Is Not as Strong as It Appears," Information Technology & Innovation Foundation, June 2018, http://www2.itif.org/2018-us-business-rd.pdf.

81 *"unfettered political, cultural, and intellectual power"*: Naomi Klein, *This Changes Everything* (New York: Simon & Schuster, 2014), 18.

81 *allowed elites to accrue nearly all of the economic gains since 1980*: Thomas Piketty, Emmanuel Saez, and Gabriel Zucman, "Distributional National Accounts: Methods and Estimates for the United States," *Quarterly Journal of Economics* 133, no. 2 (May 2018): 553–609, https://doi.org/10.1093/qje/qjx043.

81 *one-third of the growth in income inequality since 1972*: Jake Rosenfeld, Patrick Denice, and Jennifer Laird, "Union Decline Lowers Wages of Nonunion Workers," Economic Policy Institute, August 30, 2016, https://www.epi.org/publication/union-decline-lowers-wages-of-nonunion-workers-the-overlooked-reason-why-wages-are-stuck-and-inequality-is-growing/.

81 *when a politician tells them "saving the climate" means losing their jobs*: Susan Milligan, "Stretched Thin," *US News & World Report*, January 11, 2019, https://www.usnews.com/news/the-report/articles/2019-01-11/stretched-thin-majority-of-americans-live-paycheck-to-paycheck.

82 *Thousands of other bills included provisions*: Rob O'Dell and Nick Penzenstadler, "You Elected Them to Write New Laws. They're Letting Corporations Do It Instead," Center for Public Integrity, April 4, 2019, https://publicintegrity.org/politics/state-politics/copy-paste-legislate/you-elected-them-to-write-new-laws-theyre-letting-corporations-do-it-instead/.

82 *"will of the local voters and their elected leaders"*: Ibid.

83 *"total war effort"*: Ralph J. Watkins, "Economic Mobilization," *American Political Science Review* 43, no. 3 (June 1949): 555–63, https://www.jstor.org/stable/1950076?read-now=1&seq=1#page_scan_tab_contents.

83 *Buildings produce 40 percent of our nation's carbon dioxide*: "Buildings and

Built Infrastructure," Environmental and Energy Study Institute, https://www.eesi.org/topics/built-infrastructure/description.

83 *95 percent of all energy from wind and solar*: Mark Z. Jacobson, Mark A. Delucchi, Mary A. Cameron, and Bethany A. Frew, "Low-Cost Solution to the Grid Reliability Problem with 100% Penetration of Intermittent Wind, Water, and Solar for All Purposes," *Proceedings of the National Academy of Sciences* 112, no. 49 (2015): 15060–15065, https://doi.org/10.1073/pnas.1510028112.

83 *A utility-scale wind turbine*: "Wind Manufacturing and Supply Chain," Office of Energy Efficiency and Renewable Energy, https://www.energy.gov/eere/wind/wind-manufacturing-and-supply-chain.

85 *125,000 planes, 75,000 tanks, and 35,000 anti-aircraft guns*: Franklin D. Roosevelt, "The Annual Message to Congress," The White House, January 6, 1942, https://web.viu.ca/davies/H324War/FDR.message.Congress.Jan6.1942.htm.

85 *0.3 million deadweight tons*: Mark R. Wilson, *Destructive Creation: American Business and the Winning of World War II* (Philadelphia: University of Pennsylvania Press, 2016), p. 79.

85 *100 tanks and 3,700*: Harry C. Thompson and Lida Mayo, *The Ordnance Department: Procurement and Supply* (Washington, DC: Center of Military History, U. Army, 1991); Irving Brinton Holley, Jr., *Buying Aircraft: Matériel Procurement for the Army Air Forces*, Office of the Chief of Military History, Department of the Army, 1964; L. S. Ness, *Jane's World War II Tanks and Fighting Vehicles: The Complete Guide* (New York: HarperCollins, 2002), 184; *Army Air Force Statistics Digest World War II*, US Archives, December 1945, 127, https://archive.org/details/ArmyAirForcesStatisticalDigestWorldWarII.

85 *18 million deadweight tons*: Mark R. Wilson, *Destructive Creation: American Business and the Winning of World War II* (Philadelphia: University of Pennsylvania Press, 2016), p. 79.

85 *787.1 million pounds*: Ibid.

85 *299,293 aircraft and 88,410 tanks*: David M. Kennedy, *Freedom from Fear: the American People in Depression and War, 1929–1945* (New York: Oxford University Press, 1999), 655.

85 *from 7 times to 288 times the prewar average*: Wilson, *Destructive Creation*, 79.

85 *Unemployment plummeted*: "Historical Statistics of the United States," Census Library, http://www2.census.gov/prod2/statcomp/documents/HistoricalStatisticsoftheUnitedStates1789-1945.pdf.

85 *US consumer economy*: Wilson, *Destructive Creation.*

86 *publicly owned factories run by privately owned businesses*: Wilson, *Destructive Creation*, 61.

86 *"close to a quarter of the nominal value of all of the nation's factories"*: Ibid., 62.

86 *Not only did business owners convert civilian factories*: Bill McKibben, "We're Under Attack from Climate Change—and Our Only Hope Is to Mobilize Like We Did in WWII," *New Republic*, August 15, 2016, https://newrepublic.com/article/135684/declare-war-climate-change-mobilize-wwii.

86 *Perhaps most importantly, civilian contractors worked with military*: Wilson, *Destructive Creation.*

86 *The US government also paid private firms*: Ibid.

87 *During WWII*: Doris Goodwin, "The Way We Won: America's Economic Breakthrough During World War II," *American Prospect*, December 19, 2001, https://prospect.org/health/way-won-america-s-economic-breakthrough-world-war-ii/.

87 *the share of the national income*: Thomas Piketty and Emmanuel Saez, "Income Inequality in the United States, 1913–2002," November 2004, https://eml.berkeley.edu/~saez/piketty-saezOUP04US.pdf.

87 *green light*: Mazzucato, *The Entrepreneurial State.*

88 *about 57 percent of all residential rooftops*: Mark Jacobson et al., "100% clean and renewable wind, water, and sunlight (WWS) all-sector energy roadmaps for the 50 United States," *Energy Environ. Sci.* 8 (2015): 2093.

88 *average-sized residential solar installation*: Sara Matasci, "How Much Do Solar Panels Cost in the US in 2020?" *EnergySage*, March 2, 2020, https://news.energysage.com/how-much-does-the-average-solar-panel-installation-cost-in-the-u-s/.

88 *but to "soft costs"*: Galen Barboose and Naïm Darghouth, "Tracking the Sun: Pricing and Design Trends for Distributed Photovoltaic Systems in the United States," Lawrence Berkeley National Laboratory, October 2019, https://emp.lbl.gov/sites/default/files/tracking_the_sun_2019_report.pdf.

88 *navigating different layers of permitting*: Jesse Burkhardt et al., "How Much Do Local Regulations Matter?: Exploring the Impact of Permitting and Local Regulatory Processes on PV Prices in the United States," Lawrence Berkeley National Laboratory, September 2014, https://emp.lbl.gov/publications/how-much-do-local-regulations-matter.

88 *new construction to be zero-carbon*: Eric O'Shaughnessy, Gregory F. Nemet, Jacquelyn Pless, and Robert Margolis, "Addressing the soft cost challenge in U.S. small-scale solar PV system pricing," *Energy Policy* 134 (2019).

89 *mobilizations distribute wealth far more effectively*: J. W. Mason, "Lessons from World War II for the Green New Deal," Roosevelt Institute, 2020. Forthcoming paper.

89 *How can we attract enough workers*: "The Cost of Child Care in Washington, DC," Economic Policy Institute, July 2019, https://www.epi.org/child-care-costs-in-the-united-states/#/DC.

90 *requires that marginalized communities*: S. 6599, New York State Senate, June 18, 2019, https://legislation.nysenate.gov/pdf/bills/2019/S6599.

91 *Maine's Green New Deal and Los Angeles's Green New Deal*: An Act to Establish a Green New Deal for Maine, Maine Legislature, http://www.mainelegislature.org/legis/bills/bills_129th/billtexts/HP092401.asp; Eric Garcetti, "L.A.'s Green New Deal," LAmayor.org, 2019, https://plan.lamayor.org/sites/default/files/pLAn_2019_final.pdf.

91 *By altering the city's building code*: "Council to Vote on Climate Mobilization Act Ahead of Earth Day," New York City Council, April 18, 2019, https://council.nyc.gov/press/2019/04/18/1730/.

91 *Senator Elizabeth Warren would invest $10 trillion*: "Leading in Green Manufacturing," Elizabethwarren.com, June 4, 2019, https://elizabethwarren.com/plans/green-manufacturing.

91 *$15.5 billion toward sustainable agriculture*: "A New Farm Economy," Elizabethwarren.com, August 7, 2019, https://elizabethwarren.com/plans/new-farm-economy.

91 *$1 trillion to frontline*: "Fighting for Justice as We Combat the Climate Crisis," Elizabethwarren.com, https://elizabethwarren.com/plans/environmental-justice.

91 *roughly $2.7 trillion to help working-class families*: "The Green New Deal," Berniesanders.com, https://berniesanders.com/en/issues/green-new-deal/.

92 *reliable backup power . . . wetland restoration*: Ibid.

92 *"plan, implement, and administer"*: Alexandria Ocasio-Cortez, H.R. 109, February 7, 2019, https://www.congress.gov/bill/116th-congress/house-resolution/109/text.

92 *Climate Leadership and Community Protection Act*: S. 6599. New York State Senate, June 18, 2019, https://legislation.nysenate.gov/pdf/bills/2019/S6599.

92 *Similarly, at the federal level, the Climate Equity Act*: Kamala Harris and Alexandria Ocasio-Cortez, "To Ensure Climate and Environmental Justice Accountability, and for Other Purposes," 2019, https://www.harris.senate.gov/imo/media/doc/DISCUSSION%20DRAFT%20-%20Climate%20Equity%20Act.pdf.

CHAPTER 7: *THE ECONOMIC CASE FOR A GREEN NEW DEAL* / JOSEPH STIGLITZ

95 *not releasing it in the first place*: Renee Cho, "Can Removing Carbon from the Atmosphere Save Us from Climate Catastrophe?," *Columbia University Earth Institute State of the Planet* (blog), November 27, 2018, https://blogs.ei.columbia.edu/2018/11/27/carbon-dioxide-removal-climate-change/; "The True Cost of Carbon Pollution," Environmental Defense Fund, https://www.edf.org/true-cost-carbon-pollution; Kenneth Gillingham and James. H. Stock, "The Cost of Reducing Greenhouse Gas Emissions," *Journal of Economic Perspectives* 32, no. 4 (Fall 2018): 53–72, https://www.aeaweb.org/articles?id=10.1257/jep.32.4.53.

The IPCC considers soil carbon sequestration to be capable of reducing CO_2 at a low cost ($0–$100 per ton). It is estimated that it could remove 2–5 gigatons of CO_2 a year by 2050. In comparison, the world's power plants released 32.5 gigatons in 2017. Reforestation may be cheaper, at $0–$20 per ton, but has to compete for land use for agriculture, as food production needs to increase 70 percent by 2050 to feed a growing world population. Other measures are more costly. Costs for bioenergy with carbon capture and storage, for example, range from $30 to $400 per ton, and the IPCC estimates that it could remove only 0.5–5 gigatons a year by 2050. Direct air capture has become less costly (from an initial $600 per ton), but it still costs $100–$200 a ton (see Cho, above). For estimates of the cost of reducing all greenhouse gas emissions (not just CO_2), see Gillingham and Stock, above.

96 *including floods, hurricanes, and forest fires*: Extreme weather, made worse by climate change, along with the health effects of burning fossil fuels, has cost the US economy at least $240 billion a year over the past ten years. See Robert Watson, James J. McCarthy, and Liliana Hisas, "The Economic Case for Climate Action in the United States," Universal Ecological Fund FEU-US, September 2017, https://feu-us.org/case-for-climate-action-us/.

97 *controlling greenhouse gas emissions to rein in climate change*: IPCC, "Summary for Policymakers," in *Global Warming of 1.5°C: An IPCC Special Report on the Impacts of Global Warming of 1.5°C Above Pre-industrial Levels and Related Global Greenhouse Gas Emission Pathways, in the Context of Strengthening the Global Response to the Threat of Climate Change, Sustainable Development, and Efforts to Eradicate Poverty*," IPCC, 2018, https://www.ipcc.ch/sr15/chapter/spm/; Jeremy Martinich and Allison Crimmins, "Climate Damages and Adaptation Potential Across Diverse Sectors of the United States," *Nature*, April 8, 2019, https://www.nature.com/articles

/s41558-019-0444-6; Matthew E. Kahn, Kamiar Mohaddes, Ryan N. C. Ng, M. Hashem Pesaran, Mehdi Raissi, and Jui-Chung Yang, "Long-Term Macroeconomic Effects of Climate Change: A Cross-Country Analysis," Federal Reserve Bank of Dallas, July 1, 2019, https://www.dallasfed.org/~/media/documents/institute/wpapers/2019/0365.pdf.

Martinich and Crimmins project that 2.8 degrees of warming compared to preindustrial levels by 2100—well within the range we are headed toward—could cost the US almost $300 billion each year. The 2018 IPCC report provided strong arguments that warming above 1.5°C should be viewed as dangerous. Kahn et al., based on cross-country analysis, suggest that in the absence of mitigation policies, a persistent increase in average global temperature by 0.04 degrees per year would reduce world real GDP per capita by 7.22 percent by 2100.

97 *extreme cold spells during the winter of 2019*: "Late January 2019 Extreme Cold Survey," National Weather Service, https://www.weather.gov/fgf/2019_01_29-31_ExtremeCold.

97 *uncounted number of lives lost*: Y. T. Eunice Lo et al., "Increasing Mitigation Ambition to Meet the Paris Agreement's Temperature Goal Avoids Substantial Heat-Related Mortality in US Cities," *Science Advances* 5, no. 6 (2019), https://doi.org/10.1126/sciadv.aau4373; "The Economic Consequences of Climate Change," Organisation for Economic Co-operation and Development, November 3, 2015, https://doi.org/10.1787/9789264235410-en; "Climate Change and Health," World Health Organization, February 1, 2018, https://www.who.int/news-room/fact-sheets/detail/climate-change-and-health; Tamma Carleton et al., "Valuing the Global Mortality Consequences of Climate Change Accounting for Adaptation Costs and Benefits," University of Chicago Becker Friedman Institute for Economics Working Paper No. 2018-51, July 31, 2019, rev. August 12, 2019, http://dx.doi.org/10.2139/ssrn.3224365; Philip J. Landrigan et al., "Pollution and Global Health—An Agenda for Prevention," *Environmental Health Perspectives*, August 6, 2018, https://doi.org/10.1289/EHP3141; Andrew L. Goodkind, Christopher W. Tessum, Jay S. Coggins, Jason D. Hill, and Julian D. Marshall, "Fine-Scale Damage Estimates of Particulate Matter Air Pollution Reveal Opportunities for Location-Specific Mitigation of Emissions," *Proceedings of the National Academy of Sciences of the United States of America* 116, no. 18 (April 30, 2019), first published April 8, 2019, https://doi.org/10.1073/pnas.1816102116.

Consider just one aspect of the health costs—heat-related deaths. Lo et al., focusing on fifteen US cities where reliable climate and health data is available, find that achieving the 2°C threshold could save up to 1,980 annual heat-related deaths per city during extreme events, and that achieving 1.5°C could avoid up to 2,270 heat-related deaths. Globally, the OECD used the standard statistical life approach to value the economic costs of heat-related deaths; it estimated that these would increase from $100 billion today to $320 billion in 2030 and $670 billion in 2050, with the highest cost in Europe and North America (see Organisation for Economic Co-operation and Development, above).

WHO has provided a broader estimate of the cost to health: Between 2030 and 2050, climate change is expected to cause approximately 250,000 additional deaths per year, from malnutrition, malaria, diarrhea, and heat stress, with the associated direct-damage costs to health estimated to be $2–4 billion/year by 2030 (see World Health Organization, above). These numbers are perhaps conservative compared to those of a University of Chicago study, which concluded that "the cost of climate-driven changes in global mortality risk alone are as large as previous estimates of the economy-wide toll of climate change, according to an analysis of mortality data from 56% of the world's population . . . [T]he increased global mortality burden from climate change [is estimated] to be 3.7% of global GDP by the end of the century if past emissions trends continue . . . Even after accounting for adaptation, an additional 1.5 million people die per year from climate change by 2100 if past emissions trends continue" (see Carleton et al., above). There are further health costs that arise not from the increased greenhouse gases and the subsequent effects on climate but from the lethal air pollution that accompanies the burning of fossil fuels. This is associated with several million deaths per year around the world (see Landrigan, above). Similarly, Goodkind et al. estimate that more than 100,000 Americans die each year from illnesses caused by air pollution, most of which is related to the use of fossil fuels, especially coal.

97 *become inhospitable to agriculture and people*: Kanta K. Rigaud et al., "Groundswell: Preparing for Internal Climate Migration, Vol. 2: Main Report," World Bank, March 18, 2018, http://documents.worldbank.org/curated/en/846391522306665751/Main-report. Rigaud et al. show that standard estimates put the number of climate refugees by 2050 at almost 150 million.

98 *increase in temperature of 3.5 degrees Celsius*: William Nordhaus, "Projections and Uncertainties About Climate Change in an Era of Minimal Climate

Policies," *American Economic Journal: Economic Policy* 10, no. 3 (August 2018), 333–60, https://doi.org/10.1257/pol.20170046.

98 *agreed on in the Paris and Copenhagen accords*: Paris Agreement, United Nations, December 12, 2015, https://unfccc.int/files/essential_background/convention/application/pdf/english_paris_agreement.pdf; Copenhagen Accord, United Nations, December 18, 2009, https://unfccc.int/resource/docs/2009/cop15/eng/l07.pdf.

98 *carbon-heavy atmosphere not seen for more than 3 million years*: Jessica Blunden and Derek S. Arndt, "Atmospheric Composition," in *State of the Climate in 2018, Bulletin of the American Meteorological Society*, September 27, 2019, https://doi.org/10.1175/2019BAMSStateoftheClimate.1; Dieter Lüthi et al., "High-Resolution Carbon Dioxide Concentration Record 650,000–800,000 Years Before Present," *Nature*, May 15, 2008, https://doi.org/10.1038/nature06949.

98 *nonlinear effects*: Naomi Oreskes and Nicholas Stern, "Climate Change Will Cost Us Even More Than We Think," *New York Times*, October 23, 2019, https://www.nytimes.com/2019/10/23/opinion/climate-change-costs.html.

The consequences (both in terms of incidence of extreme weather events and their costs) increase much more than in proportion to the increase in greenhouse gases and temperature. "Cascading effects" are an example of such nonlinearities.

98 *the "fat tail" of unmitigated climate change*: The late Martin Weitzman persuasively argued that the probability distributions have "fat tails"—that is, extreme bad outcomes are far more likely than would arise with a normal distribution. See Martin L. Weitzman, "Fat-Tailed Uncertainty in the Economics of Catastrophic Climate Change," *Review of Environmental Economics and Policy* 5, no. 2 (Summer 2011): 275–92, https://scholar.harvard.edu/files/weitzman/files/fattaileduncertaintyeconomics.pdf.

99 *regions of the earth made uninhabitable by direct heat*: David Wallace-Wells, *The Uninhabitable Earth* (New York: Tim Duggan Books, 2019), https://www.penguinrandomhouse.com/books/586541/the-uninhabitable-earth-by-david-wallace-wells/.

Typically, those who argue for "wait and see"—maintaining that it's a waste of money to take large actions today for an uncertain risk some time far into the future—typically "discount" these future risks at a high rate. That is, whenever one takes an action that has a future cost or benefit, one has to assess the current value of these future costs or benefits. If $1 fifty years from now is worth the same as $1 today, one might be motivated to take strong action to prevent a loss; but if $1 fifty years from now is worth 3 cents, one wouldn't.

The discount rate (how we value future costs and benefits relative to today) thus becomes critical. The Trump administration has in fact said that one wouldn't want to spend more than roughly 3 cents today to prevent a $1 loss in fifty years—future generations just don't count much. And this is in line with some of the economists who are willing to accept a change in temperature of 3 to 3.5°C. The critical fallacy in their reasoning is that they look at the return to capital, some estimates of which may be as high as 7 percent. If that is the case, then *if* one took 3 cents today and put it into an insurance fund, in fifty years one would have $1, enough to offset the anticipated $1 loss. Moreover, the advocates of doing nothing then argue that this *intertemporal trade-off* (the seemingly observed rate of return) reflects society's intertemporal preferences, what it is *willing* to give up today for a dollar in the future, giving normative significance to the 7 percent rate of discount.

There are so many things wrong with this chain of reasoning that it's hard to know where to begin: it's obvious that we aren't (and haven't been) putting money aside for the potential future calamity; in fact, if anything, both public and private investment (relative to GDP) has been in decline. The 7 percent discount rate would only have significance from a normative perspective in a perfect market, where there was full intergenerational redistribution reflecting society's concerns for the future; again, it should be obvious that that is not the case. The 7 percent discount rate reflects the pervasive presence of risk, and the appropriate response to risk is *not* to increase the discount rate. Doing so confuses how we value the future with how we assess risk. Indeed, the presence of risk may (as we have already noted) lead us to want to put aside *more* today for such future contingencies.

99 *disruption of communities*: Ruth DeFries et al., "The Missing Economic Risks in Assessments of Climate Change Impacts," Policy Publication by Grantham Research Institute on Climate Change and the Environment, London School of Economics and Political Science, September 20, 2019, http://www.lse.ac.uk/GranthamInstitute/publication/the-missing-economic-risks-in-assessments-of-climate-change-impacts/.

Oreskes and Stern, in their opinion piece (see above), bring up this example from DeFries et al.: "the melting of Himalayan glaciers and snow will profoundly affect the water supply of communities in which hundreds of millions of people live, yet this is absent from most economic assessments."

100 *a savings glut*: A savings glut was argued (defensively, to help explain the

financial crisis, to which the Fed had contributed as a result of the deficiencies in its regulations) by former chairman of the Federal Reserve Ben Bernanke. Even ignoring the need for investments for retrofitting the economy to meet the challenges of climate change, Bernanke's claim that there was a savings glut seemed odd: the US needed huge investments, not only for climate change but for basic infrastructure, education, and research. Needs for investment were even greater in developing countries and emerging markets.

102 *green transition of the United States and the world*: There are already several state green banks, including an active one in New York—"NY Green Bank: Agent for Greater Private Sector Investment in Sustainable Infrastructure," New York State Energy Research and Development Authority, https://greenbank.ny.gov/About/About.

102 *more women to participate in the labor force*: Claudia Goldin, "The Quiet Revolution That Transformed Women's Employment, Education, and Family," *American Economic Review* 96, no. 2 (May 2006): 1–21, https://www.aeaweb.org/articles?id=10.1257/000282806777212350; National Research Council, *Aging and the Macroeconomy: Long-Term Implications of an Older Population* (Washington, DC: National Academies Press, 2012).

103 *highly profitable corporations to get away with paying almost no taxes*: In 2018, sixty Fortune 500 companies, some of the largest and most profitable firms, paid no federal income taxes. See Matthew Gardner, Steve Wamhoff, Mary Martellotta, and Lorena Roque, "Corporate Tax Avoidance Remains Rampant Under New Tax Law," Institute on Taxation and Economic Policy, April 11, 2019, https://itep.org/notadime/; "Key Elements of the US Tax System," in *The Tax Policy Center Briefing Book: A Citizen's Guide to the Tax System and Tax Policy* (Urban Institute and Brookings Institution), https://www.taxpolicycenter.org/briefing-book/how-do-us-corporate-income-tax-rates-and-revenues-comp.

103 *be spent to fight climate change*: Tucker Higgins, "Elizabeth Warren Would Double Her Proposed Billionaire Wealth Tax to Help Fund 'Medicare for All,'" CNBC, November 1, 2019, https://www.cnbc.com/2019/11/01/elizabeth-warren-would-double-billionaire-wealth-tax-for-medicare-for-all-plan.html; Avinash Persaud, "The Economic Consequences of the EU Proposal for a Financial Transaction Tax," *Intelligence Capital*, March 2012, https://www.robinhoodtax.org.uk/sites/default/files/The%20Economic%20Consequences%20of%20the%20EU%20Proposal%20for%20a%20Financial%20Transaction%20Tax.pdf; Joseph Stiglitz, "Reforming Taxation to Promote Growth and Equity," Roosevelt Institute White Paper,

May 28, 2014, https://rooseveltinstitute.org/reforming-taxation-promote-growth-and-equity/; Lily L. Batchelder and David Kamin, "Taxing the Rich: Issues and Options," The Aspen Institute Economic Strategy Group, September 18, 2019, https://papers.ssrn.com/sol3/papers.cfm?abstract_id=3452274.

Elizabeth Warren's proposal of a 2 percent wealth tax on wealth over $50 million and an additional 14 percent tax on wealth over a billion dollars in the US is estimated to raise $1 trillion over the next ten years (see Higgins, above). A financial transactions tax at the rate of 0.01 percent across the EU was estimated to raise more than £48 billion per year (see Persaud, above) and would raise even more in the US. In 2014, I estimated that just closing the loopholes in capital taxation and taxing capital fairly (at the same rate as labor) in the US would raise some $4 trillion over ten years (see Stiglitz, above). That number is, by now, much larger. Similarly, Batchelder and Kamin calculate that a repeal of the Tax Cuts and Jobs Act, along with additional measures, such as eliminating accelerated cost recovery for the largest businesses, taxing accrued gains at death, broadening the base of the self-employment tax, and reforming estate taxes, could raise $4.4–$4.9 trillion over the decade, or 1.6–1.8 percent of US GDP.

103 *there is scope for government to undertake more debt*: Olivier Blanchard, "Public Debt and Low Interest Rates," AEA Presidential Lecture, January 2019, https://www.aeaweb.org/aea/2019conference/program/pdf/14020_paper_etZgfbDr.pdf; Carmen M. Reinhart and Kenneth S. Rogoff, "Growth in a Time of Debt," *American Economic Review* 100, no. 2 (May 2010): 573–78, http://www.aeaweb.org/articles.php?doi=10.1257/aer.100.2.573; Thomas Herndon, Michael Ash, and Robert Pollin, "Does High Public Debt Consistently Stifle Economic Growth? A Critique of Reinhart and Rogoff," *Cambridge Journal of Economics* 38, no. 2 (March 2014): 257–79, https://doi.org/10.1093/cje/bet075.

The argument put forward by Reinhardt and Rogoff—that there is a critical debt to GDP threshold, above which growth slows—has been thoroughly discredited by Herndon et al. There is a large literature, both technical and popular, explaining why debt fetishism is foolish.

104 *shrunk to around 55 percent by 1955 and below 40 percent by 1964*: "Historical Data on Federal Debt Held by the Public," Congressional Budget Office Economic and Budget Issue Brief, August 5, 2010, https://www.cbo.gov/publication/21728.; Valerie Ramy, "It's Time to Start Worrying About the National Debt," *Wall Street Journal*, August 23, 2019; for treatment of

debt-to-GDP ratio generally in Europe and the US see Joseph E. Stiglitz, *The Euro: How a Common Currency Threatens the Future of Europe* (New York: W. W. Norton & Company, 2016).

104 *marginalized groups have been effectively brought into the labor force*: For instance, analysis of BLS CPS data shows that the unemployment rate of young African Americans fell to a low of 7.6 percent between 1995 and 2000—in contrast to what happened when the economy went into recession, when it soared to 16.8 percent in March 2010. See Valerie Wilson, "The Impact of Full Employment on African American Employment and Wages," Economic Policy Institute, March 20, 2015, https://www.epi.org/publication/the-impact-of-full-employment-on-african-american-employment-and-wages/; US Bureau of Labor Statistics, "Current Population Survey," https://www.census.gov/programs-surveys/cps.html.

105 *already spurred a lot of innovation*: "What Have We Learned from Attempts to Introduce Green-Growth Policies?," Organisation for Economic Co-operation and Development, March 2015, https://www.oecd-ilibrary.org/docserver/5k486rchlnxx-en.pdf?expires=1585184618&id=id&accname=guest&checksum=8B6ADAD7EEAE5D24747F1808B90648F7.

CHAPTER 8: *A GREEN NEW DEAL FOR THE GULF SOUTH* / COLETTE PICHON BATTLE

108 *across the five Gulf South states*: See the full policy platform and join our organizing initiatives at https://www.gcclp.org/gulf-south-for-a-green-new-deal.

The term "at will" means that an employer can terminate an employee for any reason without warning. However, an employer cannot fire an employee illegally—for instance, because of gender, race, or religion. Contracts in at-will states between employee and employer prevent the employee from pursuing a claim against the employer as a result of being fired. In other words, an employee cannot sue for lost wages due to dismissal from the job, provided the dismissal was legal, as discussed above. Mississippi and Texas have at-will laws, but these states do allow for a few exceptions.

In states where there are right-to-work laws, employers and labor unions are prohibited from forcing any employee who is not part of the union to pay fees. Some states also include language that prohibits employers and unions from requiring union membership as a condition of employment.

109 *American working-class fight against exploitation and violent suppression*: "NAACP History: W. E. B. Du Bois," National Association for the Advancement of Colored People, 2019, https://www.naacp.org/naacp-history-w-e-b-dubois/.

109 *remains a statement of fact today*: Unity and Struggle, "As the South Goes, So Goes the Nation," *Unity and Struggle* (blog), November 30, 2009, http://www.unityandstruggle.org/2009/11/as-the-south-goes-so-goes-the-nation/.

109 *produces more crude oil than any other state*: "Texas State Energy Profile," US Energy Information Administration, March 19, 2020, https://www.eia.gov/state/print.php?sid=TX.

109 *The Gulf Coast is home*: "GULF OF MEXICO FACT SHEET," U.S. Energy Information Administration, July 12, 2019.

109 *This toxic economy, its foundations*: Michael Isaac Stein, "What Will Happen to the Gulf Coast If the Oil Industry Retreats?," CityLab, September 20, 2017, https://www.citylab.com/environment/2017/09/what-will-happen-to-the-gulf-coast-if-the-oil-industry-retreats/540372/.

110 *"further rig the system at the expense of working families"*: "Right to Work," American Federation of Labor and Congress of Industrial Organizations, 2019, https://aflcio.org/issues/right-work.

111 *nearly 2 million homes at risk in just Florida and Texas alone*: "New Study Finds 82,000 Texas Homes Worth $17 Billion Will Be at Risk from Tidal Flooding," press release, Union of Concerned Scientists, June 18, 2018, https://www.ucsusa.org/press/2018/new-study-finds-82000-texas-homes-worth-17-billion-will-be-risk-tidal-flooding.

111 *highest relative levels of land loss to sea level rise on the planet*: "Plaquemines Parish, Louisiana: Sinking Land and Rising Seas Mean Tough Choices (2015)," *Union of Concerned Scientists* (blog), November 8, 2016, https://www.ucsusa.org/global-warming/science-and-impacts/impacts/plaquemines-parish-louisiana-sinking-land-and-rising.

111 *is expected to quadruple by 2050*: "Southeast Region Areas to Endure About Four Months a Year When 'Feels Like' Temperature Exceeds 105 Degrees," Union of Concerned Scientists, press release, July 16, 2019, https://www.ucsusa.org/press/2019/southeast-region-areas-endure-about-four-months-year-when-feels-temperature-exceeds-105.

115 *Vision for Black Lives platform*: "A Vision for Black Lives: Policy Demands for Black Power, Freedom, and Justice," The Movement for Black Lives, 2016, https://neweconomy.net/sites/default/files/resources/20160726-m4bl-Vision-Booklet-V3.pdf.

CHAPTER 9: *GREEN NEW BINGO HALL* / JULIAN BRAVE NOISECAT

120 *10 million barrels per day, according to the International Energy Agency*: "EIA: US Overtakes Russia, Saudi Arabia as World's Largest Crude

Producer," World Oil, September 12, 2018, https://www.worldoil.com/news/2018/9/12/eia-us-overtakes-russia-saudi-arabia-as-worlds-largest-crude-producer.

120 *concentration of carbon dioxide in the air reached 415 parts per million*: "Daily CO_2," CO_2-Earth, February 23, 2020, https://www.co2.earth/daily-co2.

121 *United States provides the fossil fuel industry over $500 billion*: David Coady, Ian Parry, Nghia-Piotr Le, and Baoping Shang, "Global Fossil Fuel Subsidies Remain Large: An Update Based on Country-Level Estimates," International Monetary Fund, May 2, 2019, https://www.imf.org/en/Publications/WP/Issues/2019/05/02/Global-Fossil-Fuel-Subsidies-Remain-Large-An-Update-Based-on-Country-Level-Estimates-46509.

121 *We are a $20 trillion economy*: "United States GDP," Trading Economics, 2020, https://tradingeconomics.com/united-states/gdp.

121 *we have spewed more carbon into the atmosphere than in all prior human history*: David Wallace-Wells, *The Uninhabitable Earth: Life after Warming* (New York: Tim Duggan Books, 2019), 4; Peter Frumhoff, "Global Warming Fact: More Than Half of All Industrial CO_2 Pollution Has Been Emitted Since 1988," *Union of Concerned Scientists* (blog), December 15, 2014, https://blog.ucsusa.org/peter-frumhoff/global-warming-fact-co2-emissions-since-1988-764.

121 *After three years of decline, US emissions went up 3.4 percent in 2018*: Kris Maher, "US Carbon Emissions Rose 3.4% in 2018 as Economy Surged," *Wall Street Journal*, January 8, 2019, https://www.wsj.com/articles/u-s-carbon-emissions-rose-3-4-in-2018-as-economy-surged-11546978889.

121 *described global warming as "a hoax"; "the golden era of American energy"*: Louis Jacobson, "Yes, Donald Trump Did Call Climate Change a Chinese Hoax," PolitiFact, June 3, 2016, https://www.politifact.com/factchecks/2016/jun/03/hillary-clinton/yes-donald-trump-did-call-climate-change-chinese-h/; Donald J. Trump, The White House, May 14, 2019, https://www.whitehouse.gov/briefings-statements/president-donald-j-trump-unleashing-american-energy-dominance/.

122 *would contribute 24.3 million metric tons to CO_2 emissions*: "Climate Impacts of the Keystone XL Tar Sands Pipeline," Natural Resources Defense Council, October 2013, https://assets.nrdc.org/sites/default/files/tar-sands-climate-impacts-IB.pdf.

123 *In 2016, more than 1 million people checked in at Standing Rock*: Merrit Kennedy, "More Than 1 Million 'Check In' on Facebook to Support the Standing Rock Sioux," NPR, November 1, 2016, https://www.npr.org/sections

/thetwo-way/2016/11/01/500268879/more-than-a-million-check-in-on-facebook-to-support-the-standing-rock-sioux.

123 *One hundred six members of Congress have co-sponsored a Green New Deal resolution*: Recognizing the Duty of the Federal Government to Create a Green New Deal, H.R. 109, https://www.congress.gov/bill/116th-congress/house-resolution/109/cosponsors?searchResultViewType=expanded&KWICView=false; A Resolution Recognizing the Duty of the Federal Government to Create a Green New Deal, S.R. 59, https://www.congress.gov/bill/116th-congress/senate-resolution/59/cosponsors?searchResultViewType=expanded&KWICView=false.

123 *eighteen presidential candidates endorsed the vision*: "Green New Deal: Candidate Scorecards," Data for Progress, 2019, https://www.dataforprogress.org/gnd-candidates.

123 *Climate Equity Act*: "Harris, Ocasio-Cortez Announce Landmark Legislation to Ensure Green New Deal Lifts Up Every Community," press release, Harris.senate.gov, July 29, 2019, https://www.harris.senate.gov/news/press-releases/harris-ocasio-cortez-announce-landmark-legislation-to-ensure-green-new-deal-lifts-up-every-community.

123 *Climate Leadership and Community Protection Act*: S. 6599, New York State Senate, https://www.nysenate.gov/legislation/bills/2019/s6599.

123 *CNN's town hall on the climate crisis*: Steven Collinson, "What Happened During CNN's Climate Town Hall and What It Means for 2020," CNN, September 5, 2019, https://www.cnn.com/2019/09/05/politics/climate-town-hall-highlights/index.html.

125 *the intergenerational memory and experience of a people who survived a genocide*: Julian Brave NoiseCat, "We Need Indigenous Wisdom to Survive the Apocalypse," *The Walrus*, March 27, 2020, https://thewalrus.ca/we-need-indigenous-wisdom-to-survive-the-apocalypse/.

125 *the end of the world*: Julian Brave NoiseCat, "Perhaps the World Ends Here," *Harper's*, December 5, 2019, https://harpers.org/blog/2019/12/wounded-knee-pine-ridge-reservation-perhaps-the-world-ends-here/.

CHAPTER 10: *A WORKERS' GREEN NEW DEAL* / MARY KAY HENRY

127 *lives of working families*: Sharon Block, "How Labor Law Could Help—Not Hinder—Tackling Big Problems," *OnLabor* (blog), December 6, 2019, https://onlabor.org/how-labor-law-could-help-not-hinder-tackling-big-problems/.

127 *service and caregiving sectors*: Congressional Research Service, Table A1: "Union

Membership in the United States, 1930–2003," Union Membership Trends in the United States, Cornell University, 2004, CPS-22, https://digitalcommons.ilr.cornell.edu/cgi/viewcontent.cgi?article=1176&context=key_workplace.

128 *ability to organize unions*: Lisa Graves, "Inside the Koch Family's 60-Year Anti-Union Campaign That Gave Us Janus," *In These Times*, June 19, 2018, https://inthesetimes.com/working/entry/21294/koch_anti_union_janus_supreme_court.

128 *preserve dirty energy sources*: "Koch Industries: Still Fueling Climate Denial," Greenpeace, April 14, 2011, https://www.greenpeace.org/usa/research/koch-industries-still-fueling-climate-denial-2011-update/.

129 *better than his or her parents*: Raksha Kopparam, "To Fight Falling US Intergenerational Mobility, Tackle Economic Inequality," *Washington Center for Equitable Growth* (blog), December 5, 2019, https://equitablegrowth.org/to-fight-falling-u-s-intergenerational-mobility-tackle-economic-inequality/?mkt_tok=eyJpIjoiWWpkbE5UbGtNV0k1WW1KaSIsInQiOiJmNXdGSW43WG54WFVkd2RSZnhVSDFVcmJHWWRKdVNpU0VVUTcrbmE1dXVVdUVRNm90bHhBeFdFRFBCT2ZDWDM4aHR2SjcyclRGelJSaWt0bkRLQjQ4OUhxOXVkXC9iWWcxUUUyV1kwWHlRTGZKRTRCTmw2Unc4cVlcL0hnckMrU2R4In0%3D.

130 *eleven years after Katrina*: Gary Rivlin, "White New Orleans Has Recovered from Hurricane Katrina. Black New Orleans Has Not," *The Nation*, August 26, 2016, https://www.thenation.com/article/archive/white-new-orleans-has-recovered-from-hurricane-katrina-black-new-orleans-has-not/.

130 *over a decade later*: "Neighborhood Change Rates: Growth Continues Through 2018," The Data Center, August 23, 2018, https://www.datacenterresearch.org/reports_analysis/neighborhood-recovery-rates-growth-continues-through-2018-in-new-orleans-neighborhoods/.

131 *By 1953, nearly 35 percent*: Gerald Mayer, "Union Membership Trends in the United States," Congressional Research Service, 2004.

131 *11 percent of working Americans belong to a union*: US Bureau of Labor Statistics, "Union Members Summary," January 22, 2020, https://www.bls.gov/news.release/union2.nr0.htm.

132 *45 percent of American workers are now effectively*: "Estimates of Workers Excluded from Collective Bargaining," Service Employees International Union, Strategic Initiatives internal memo.

134 *"$15 and a Union" movement*: Alana Semuels, "Fast-Food Workers Walk Out in N.Y. Amid Rising U.S. Labor Unrest," *Los Angeles Times*, November 29, 2012, https://www.latimes.com/business/la-xpm-2012-nov-29-la-fi

-mo-fast-food-strike-20121129-story.html; "Our Role," New York Communities for Change, Brooklyn, NY, n.d., https://www.nycommunities.org/issues/labor/.

CHAPTER 11: *PEOPLE POWER AND POLITICAL POWER* / VARSHINI PRAKASH

138 *a representative of the youth-led Energy Action Coalition*: Suzanne Goldenberg, "Revealed: The Day Obama Chose a Strategy of Silence on Climate Change," *Mother Jones*, November 3, 2012, https://www.motherjones.com/environment/2012/11/obama-climate-change-silence/2/.

139 *stopped 132 coal plants*: Mark Hertsgaard, "How a Grassroots Rebellion Won the Nation's Biggest Climate Victory," *Mother Jones*, April 2, 2012, https://www.motherjones.com/environment/2012/04/beyond-coal-plant-activism/2/.

139 *bans and moratoriums on shale gas extraction*: Liz Edmondson, "Regulating Hydraulic Fracturing in the States: Trending Issues in 2016 and Beyond," Council of State Governments, July 1, 2016, https://knowledgecenter.csg.org/kc/content/regulating-hydraulic-fracturing-states-trending-issues-2016-and-beyond.

140 *"divestment efforts affecting the investment community"*: Form 10-K, Annual Report Pursuant to Section 13 or 15(d) of the Securities Exchange Act of 1934, for the Fiscal Year Ended December 31, 2015, https://www.sec.gov/Archives/edgar/data/1064728/000106472816000157/btu-20151231x10k.htm.

141 *and repealing countless climate and environmental regulations*: Nadja Popovich, Livia Albeck-Ripka, and Kendra Pierre-Louis, "95 Environmental Rules Being Rolled Back Under Trump," *New York Times*, December 21, 2019, https://www.nytimes.com/interactive/2019/climate/trump-environment-rollbacks.html.

145 *"Gay marriage isn't winning the day because of some singularly persuasive legal argument"*: Richard Kim, "Why Gay Marriage Is Winning," *The Nation*, July 13, 2013, https://www.thenation.com/article/why-gay-marriage-won/.

145 *Only 32 percent*: Mark Engler and Paul Engler, "This Is an Uprising: How Nonviolent Revolt Is Shaping the Twenty-First Century," (New York: PublicAffairs, 2016), 222.

145 *almost seven in ten Americans*: Center for American Progress Immigration Team, "The Facts on Immigration Today," July 6, 2012, https://www.americanprogress.org/issues/immigration/reports/2012/07/06/11888/the-facts-on-immigration-today/.

146 *an army of antitax lobbyists have managed to repeatedly lower taxes for corporations and the rich*: Ben White, "Soak the Rich? Americans Say Go for It," *Politico*, February 4, 2019, https://www.politico.com/story/2019/02/04/democrats-taxes-economy-policy-2020-1144874.

146 *campaigned steadily to defeat anti-gun politicians and legislation*: Matt Deitsch and Andrew Mangan, "Memo: Gun Violence Prevention," Data for Progress, October 24, 2019, https://www.dataforprogress.org/memos/gun-violence-prevention.

147 *"utilities [outspent] environmental groups and the renewable energy industry"*: "Fossil Fuel Interests Have Outspent Environmental Advocates 10:1 on Climate Lobbying," *Yale Environment 360*, July 19, 2018, https://e360.yale.edu/digest/fossil-fuel-interests-have-outspent-environmental-advocates-101-on-climate-lobbying.

148 *build a broad popular movement to tackle climate change*: Theda Skocpol, "Naming the Problem," The Politics of America's Fight Against Global Warming (symposium), Harvard University, January 2013, https://scholars.org/sites/scholars/files/skocpol_captrade_report_january_2013_0.pdf.

148 *Intense constituencies are levers that move politicians. Polls aren't*: David Roberts, "What Theda Skocpol Gets Right About the Cap-and-Trade Fight," *Grist*, January 15, 2013, https://grist.org/climate-energy/what-theda-skocpol-gets-right-about-the-cap-and-trade-fight/.

148 *3.5 percent*: Erica Chenoweth, "My Talk at TEDxBoulder: Civil Resistance and the '3.5% Rule,'" September 21, 2013, and undated blog post at *RationalInsurgent*, https://rationalinsurgent.com/2013/11/04/my-talk-at-tedxboulder-civil-resistance-and-the-3-5-rule/. Also see Maria J. Stephan and Erica Chenoweth, "Why Civil Resistance Works: The Strategic Logic of Nonviolent Conflict," *International Security* 33, no. 1 (Summer 2008): 7–44, https://www.jstor.org/stable/40207100, and Erica Chenoweth and Maria J. Stephan, *Why Civil Resistance Works: The Strategic Logic of Nonviolent Conflict*, Columbia Studies in Terrorism and Irregular Warfare (New York: Columbia University Press, 2011).

151 *4,000 more marched through the streets*: Martin Luther King Jr., *The Trumpet of Conscience* (Boston: Beacon Press, 2018).

151 *more than 750 civil rights demonstrations*: Taylor Branch, *Parting the Waters: America in the King Years 1954–63* (New York: Simon & Schuster, 2007), 825.

151 *unless it adopted a more radical policy*: Adam Fairclough, *To Redeem the Soul of America: The Southern Christian Leadership Conference and Martin Luther King, Jr.*, rev. ed. (Athens: University of Georgia Press, 2001), 134.

151 *"from 4 percent in early spring to 52 percent by early summer"*: Earl Black and Merle Black, *Politics and Society in the South* (Cambridge, MA: Harvard University Press, 1989), 110.

152 *"It is time to act in the Congress, in your state and local legislative body"*: "Excerpt from a Report to the American People on Civil Rights, 11 June 1963," John F. Kennedy Presidential Library and Museum, https://www.jfklibrary.org/learn/about-jfk/historic-speeches/televised-address-to-the-nation-on-civil-rights.

152 *substantially stronger than those he had proposed earlier in the year*: David J. Garrow, *Protest at Selma: Martin Luther King, Jr., and the Voting Rights Act of 1965* (New York: Open Road Media, 2015).

152 *"The Civil Rights Commission had written powerful documents . . ."*: King, *The Trumpet of Conscience*, 55.

152 *over years of struggle and hundreds of local campaigns*: See Barbara Ransby, *Ella Baker and the Black Freedom Movement: A Radical Democratic Vision* (Chapel Hill: University of North Carolina Press, 2005); Frances Fox Piven and Richard A. Cloward, *Poor People's Movements: Why They Succeed, How They Fail* (New York: Vintage, 1978); Branch, *Parting the Waters.*

152 *involved 50,000 participants across the South*: Piven and Cloward, *Poor People's Movements*, 224.

152 *years of arduous preparation, including many failures*: See Ransby, *Ella Baker and the Black Freedom Movement*; Charles M. Payne, *I've Got the Light of Freedom* (Berkeley: University of California Press, 2007).

154 *without making a fuss*: Gene Sharp, *Waging Nonviolent Struggle* (Manchester, NH: Extending Horizon Books, 2005); Erica Chenoweth and Maria J. Stephan, *Why Civil Resistance Works* (New York: Columbia University Press, 2011), Kindle Edition.

154 *workers executed the same type of noncooperation*: Jeremy Brecher, *Strike!* (Oakland, CA: PM Press, 2014); Kristin Downey, *The Woman Behind the New Deal: The Life of Frances Perkins, FDR'S Secretary of Labor and His Moral Conscience* (New York: Anchor, 2009).

154 *"shook the country" and "panicked" Roosevelt's administration*: William E. Leuchtenburg, *Franklin D. Roosevelt and the New Deal: 1932–1940* (New York: Harper Perennial, 2009), 111–14.

154 *1.5 million American workers participated in nearly 1,900 strikes:* "Analysis of Strikes and Lockouts in 1934 and Analysis for September 1935," Bureau of Labor Statistics, 1936, https://www.bls.gov/wsp/publications/annual-summaries/work-stoppages-1934-and-1935.pdf.

154 *"was indeed the consequence of popular upheaval"*: Steve Fraser, "The New Deal in the American Political Imagination," *Jacobin*, June 30, 2019, https://www.jacobinmag.com/2019/06/new-deal-great-depression.

154 *"only strong measures would restore stability"*: David Plotke, *Building a Democratic Political Order: Reshaping American Liberalism in the 1930s and 1940s* (Cambridge: Cambridge University Press, 1996), 107.

155 *"employers to accede peacefully to the unionization of their plants"*: Leuchtenburg, *Franklin D. Roosevelt and the New Deal: 1932–1940*, 151.

155 *"one in four Americans would support an organization engaging in nonviolent civil disobedience"*: Anthony Leiserowitz, Edward Maibach, Connie Roser-Renouf, and Geoff Feinberg, "How Americans Communicate About Global Warming in 2013," Yale Project on Climate Change Communication and the George Mason University Center for Climate Change Communication, April 2013, https://climatecommunication.yale.edu/wp-content/uploads/2016/02/2013_08_How-Americans-Communicate-About-Global-Warming-April-2013.pdf.

157 *produce 33 percent of electricity from renewable sources by 2027*: Brad Plumer, "Hillary Clinton Is Calling for a 700% Increase in Solar Power. Is That Realistic?," *Vox*, July 26, 2015, https://www.vox.com/2015/7/26/9044343/hillary-clinton-renewable-solar.

157 *new fossil fuel infrastructure projects like the Dakota Access Pipeline*: Coral Davenport, "Hillary Clinton's Ambitious Climate Change Plan Avoids Carbon Tax," *New York Times*, July 2, 2016, https://www.nytimes.com/2016/07/03/us/politics/hillary-clintons-ambitious-climate-change-plan-avoids-carbon-tax.htm.

157 *reduce carbon emissions 80 percent by 2050*: Alex Seitz-Wald, "Bernie Sanders Unveils Climate Plan," NBC News, December 7, 2015, https://www.nbcnews.com/politics/2016-election/bernie-sanders-unveils-climate-plan-n475366.

157 *"The Democratic Party does not have a plan to address climate change"*: Robinson Meyer, "Democrats Are Shockingly Unprepared to Fight Climate Change," *The Atlantic*, November 15, 2017, https://www.theatlantic.com/science/archive/2017/11/there-is-no-democratic-plan-to-fight-climate-change/543981/.

158 *electricity and transportation 100 percent renewable by 2030*: "Issues: The Green New Deal," Berniesanders.com, https://berniesanders.com/en/issues/green-new-deal/.

158 *carbon neutrality across the economy by 2030*: Elizabeth Warren, "100%

Clean Energy for America," *Medium*, September 3, 2019, https://medium.com/@teamwarren/100-clean-energy-for-america-de75ee39887d.

158 *a $1.7 trillion plan to achieve net-zero emissions by 2050*: "Joe's Plan for a Clean Energy Revolution and Environmental Justice," accessed February 20, 2020, https://joebiden.com/climate/.

160 *IDC challengers . . . won their primaries by just over 4,000 votes*: NY State Board of Elections, Certified Results for the September 13, 2018, Primary Election, https://www.elections.ny.gov/2018ElectionResults.html.

160 *win the endorsement of labor unions*: Kyla Mandel, "Maine's Green New Deal Bill First in Country to Be Backed by Labor Unions," Think Progress, April 16, 2019, https://thinkprogress.org/labor-union-support-maine-green-new-deal-bc048eea1c91/.

160 *Maxmin beat her Republican opponent by only 220 votes*: Tabulations for Elections Held in 2018, November 6, 2018, https://www.maine.gov/sos/cec/elec/results/results18.html#nonrcv.

161 *1,300 candidates signed the pledge*: "Pledge Signers," accessed February 22, 2020, http://nofossilfuelmoney.org/pledge-signers/.

161 *700 fossil-free candidates were on the ballot*: Ibid.

161 *Abdul ultimately fell short, winning 30 percent in a three-way race*: "Michigan Governor Primary Election Results," *New York Times*, September 24, 2018, https://www.nytimes.com/elections/results/michigan-governor-primary-election.

162 *The Democratic nominee, Gretchen Whitmer, had taken more than $10,000*: Amanda Seitz and David Eggert, "Michigan Democratic Candidates Spar over Corporate Donations," August 2, 2018, https://apnews.com/b12087a7d3cd4b7fa370db65417701f0/Michigan-Democratic-candidates-spar-over-corporate-donations.

CHAPTER 12: *WE SHINE BRIGHT: ORGANIZING IN HOPE AND SONG* / SARA BLAZEVIC, VICTORIA FERNANDEZ, DYANNA JAYE, AND ARU SHINEY-AJAY

170 *movements as living organisms through the metaphor of DNA*: Mark Engler and Paul Engler, *This Is an Uprising: How Nonviolent Revolt Is Shaping the Twenty-First Century* (New York: Nation Books, 2016), 72–76.

176 *"I'm eighteen, and I'm really concerned about the future of our country . . ."*: Dan Levin, "A Politician Called Her 'Young and Naïve.' Now She's Striking Back." *New York Times*, July 25, 2018, https://www.nytimes.com/2018/07/25/us/young-and-naive-rose-strauss.html.

179 *1.5 percent of the questions asked in the presidential primary debates*: Kevin

Kalhoefer, "Primary Debate Scorecard: Climate Change Through 20 Presidential Debates," Media Matters, March 23, 2016, https://www.mediamatters.org/donald-trump/primary-debate-scorecard-climate-change-through-20-presidential-debates.

182 *more effective at garnering support for their cause*: Ruud Wouters, "The Persuasive Power of Protest. How Protest Wins Public Support," *Social Forces* 98, no. 1 (2018): 403–26, https://academic.oup.com/sf/article-abstract/98/1/403/5158514; Ruud Wouters and Stefaan Walgrave, "Demonstrating Power: How Protest Persuades Political Representatives," *American Sociological Review* 82, no. 2 (2017), 361–83, https://www.researchgate.net/publication/315702318_Demonstrating_Power_How_Protest_Persuades_PoliticalRepresentatives.

186 *housing the Student Nonviolent Coordinating Committee (SNCC) used*: Charles Cobb, SNCC field secretary, interviewed by Terry Gross, "50 Years Ago, Students Fought for Black Rights During 'Freedom Summer,'" NPR, June 23, 2014, https://www.npr.org/transcripts/324879867.

188 *"that sense of urgency"*: Lexi McMenamin, "The Youth Climate Movement Comes to New Hampshire," *New Republic*, February 5, 2020, https://newrepublic.com/article/156476/youth-climate-movement-comes-new-hampshire.

BLUE SKIES IN AMERICA / SAYA AMELI HAJEBI

198 *over $40,000 in campaign donations*: Massachusetts Office of Campaign and Political Finance, accessed July 2018; Fossil fuel industry contributors defined by the list provided by the No Fossil Fuel Money pledge, accessed July 2018, http://nofossilfuelmoney.org/company-list/.

198 *ambitious, comprehensive climate bill*: Matt Stout and Jon Chesto, "Renewable energy could get a big boost from new bill," *Boston Globe*, July 30, 2018; Tim Cronin, "[MASSACHUSETTS] END OF SESSION CLIMATE POLICY SUMMARY," Climate Action Business Association. July 31, 2018, https://cabaus.org/2018/07/31/massachusetts-end-session-climate-policy-summary/; Christian Roselund, "Massachusetts committee produces weaker energy bill," *PV Magazine*, July 31, 2018, https://pv-magazine-usa.com/2018/07/31/massachusetts-committee-produces-weaker-energy-bill/.

CHAPTER 13: *A THIRD RECONSTRUCTION FOR OUR COMMON HOME* / REV. WILLIAM J. BARBER II

200 *"Indigenous peoples have witnessed continual ecosystem and species collapse"*: Malcolm Harris, "Indigenous Knowledge Has Been Warning Us About Climate Change for Centuries," *Pacific Standard*, March 4, 2019, https://psmag.com/ideas/indigenous-knowledge-has-been-warning-us-about-climate-change-for-centuries.

201 *the* White Lion *traded the settlers "20 and odd" Africans*: John Rolfe to Sir Edwin Sandys, January 1620, Hampton (Virginia) History Museum, text prepared by Beth Austin, Registrar and Historian, December 2018, rev. December 2019, https://hampton.gov/DocumentCenter/View/24075/1619-Virginias-First-Africans?bidId=.

203 *"the innocent daughter of Mother Nature"*: Ta-Nehisi Coates, *Between the World and Me* (New York: Spiegel & Grau, 2015), 7.

203 *"race is the child of racism, not the father"*: Ibid.

204 *"we made the world we're living in and we have to make it over again"*: James Baldwin, "Notes for a Hypothetical Novel: An Address," in *James Baldwin: Collected Essays* (New York: Library of America, 1998), 230.

205 *"beneficent provision for the poor, the unfortunate, and the orphan"*: North Carolina Constitution, Article XI, Sect. 4, https://law.justia.com/constitution/north-carolina/article_vii-xiv.html.

207 *twenty-three states have passed racist voter suppression laws*: Danny Hakim and Michael Wines, " 'They Don't Really Want Us to Vote': How Republicans Made It Harder," *New York Times*, November 3, 2018, https://www.nytimes.com/2018/11/03/us/politics/voting-suppression-elections.html.

207 *gutted Section 5 in 2013*: Adam Liptak, "Supreme Court Invalidates Key Part of Voting Rights Act," *New York Times*, June 25, 2013, https://www.nytimes.com/2013/06/26/us/supreme-court-ruling.html.

208 *can buy unleaded gas but can't get unleaded water*: Olga Khazan, "The Trouble with America's Water," *The Atlantic*, September 11, 2019, https://www.theatlantic.com/health/archive/2019/09/millions-american-homes-have-lead-water/597826/.

208 *within thirty miles of a coal-fired power plant*: Martina Jackson Haynes et al., "Coal Blooded Action Toolkit," NAACP Environmental and Climate Justice Program, https://www.naacp.org/wp-content/uploads/2016/04/Coal_Blooded_Action_Toolkit_FINAL_FINAL.pdf, 6; Clean Air Task Force,

"Air of Injustice: African Americans and Power Plant Pollution," http://www.catf.us/resources/publications/files/Air_of_Injustice.pdf.

209 *live where industrial pollution poses the greatest health danger*: Dr. Robert Bullard, "African Americans on the Frontline Fighting for Environmental Justice," *Dr. Robert Bullard* (blog), February 22, 2019, https://drrobertbullard.com/african-americans-on-the-frontline-fighting-for-environmental-justice/.

209 *near the nation's most dangerous chemical plants*: Ibid.

210 *By 2040 in the United States, white people will be one among many minority groups*: William H. Frey, "The US Will Become 'Minority White' in 2045, Census Projects," Brookings Institution, March 14, 2018, https://www.brookings.edu/blog/the-avenue/2018/03/14/the-us-will-become-minority-white-in-2045-census-projects/.

CHAPTER 14: *THE NEXT ERA OF AMERICAN POLITICS* / GUIDO GIRGENTI AND WALEED SHAHID

212 *"The green dream or whatever they call it"*: Heather Caygle, Sarah Ferris, and John Bresnahan, "'Too Hot to Handle': Pelosi Predicts GOP Won't Trigger Another Shutdown," *Politico*, February 7, 2019, https://www.politico.com/story/2019/02/07/pelosi-trump-government-shutdown-1154355?nname=playbook&nid=0000014f-1646-d88f-a1cf-5f46b7bd0000&nrid=0000014c-2414-d9dd-a5ec-34bc4cff0000&nlid=630318.

212 *"have not done the things that are necessary to lower emissions"*: Naomi Klein, *This Changes Everything* (New York: Simon & Schuster, 2014). See also pp. 27–37 of the present book.

213 *inequality higher than at any time since the Gilded Age*: Mark Schneider, "Income Inequality Grew Again: The Highest Level in More Than 50 Years, Census Bureau Says," *USA Today*, September 26, 2019, https://www.usatoday.com/story/money/2019/09/26/income-inequality-highest-over-50-years-census-bureau-shows/3772919002/.

213 *the worst economic crisis since the Great Depression*: Heather Stewart, "We Are in the Worst Financial Crisis Since Depression, Says IMF," *The Guardian*, April 9, 2008, https://www.theguardian.com/business/2008/apr/10/us economy.subprimecrisis.

213 *the longest decline in Americans' life expectancy since World War I*: "CDC Says Life Expectancy Down as More Americans Die Younger Due to Suicide and

Drug Overdose," Associated Press/CBS News, December 7, 2018, https://www.cbsnews.com/news/cdc-us-life-expectancy-declining-due-largely-to-drug-overdose-and-suicides/.

213 *the highest incarceration rate in the industrialized world*: Sintia Radu, "Countries with the Highest Incarceration Rates," *US News & World Report*, May 13, 2019, https://www.usnews.com/news/best-countries/articles/2019-05-13/10-countries-with-the-highest-incarceration-rates.

213 *the most expensive health care system*: Hristina Byrnes, "US Leads Among Countries That Spend the Most on Public Health Care," *USA Today*, April 11, 2019, https://www.usatoday.com/story/money/2019/04/11/countries-that-spend-the-most-on-public-health/39307147/.

213 *the highest levels of student debt*: Anthony Cilluffo, "5 Facts About Student Loans," Pew Research Center, August 13, 2019, https://www.pewresearch.org/fact-tank/2019/08/13/facts-about-student-loans/.

213 *the longest period without an increase in the minimum wage*: Aimee Picchi, "The Federal Minimum Wage Sets a Record—for Not Rising," CBS News, June 15, 2019, https://www.cbsnews.com/news/federal-minimum-wage-sets-record-for-length-with-no-increase/.

213 *union membership lower than at any point since the New Deal*: Steven Greenhouse, "Union Membership in US Fell to a 70-year Low Last Year," *New York Times*, January 21, 2011, https://www.nytimes.com/2011/01/22/business/22union.html.

214 *Historians and political scientists refer to these turning points as realignments*: Stephen Skowronek, *Presidential Leadership in Political Time: Reprise and Reappraisal*, 2nd ed., rev. (Lawrence: University Press of Kansas, 2011), Audible Edition; James L. Sundquist, *Dynamics of the Party System: Alignment and Realignment of Political Parties in the United States*, rev. ed. (Washington, DC: Brookings Institution, 1983); Walter Dean Burnham, *Critical Elections and the Mainsprings of American Politics* (New York: W. W. Norton & Company, 1970); C. O. Key Jr., "A Theory of Critical Elections," *Journal of Politics* 17, no. 1 (February 1955): 3–18; David R. Mayhew, *Electoral Realignments: A Critique of an American Genre* (New Haven, CT: Yale University Press, 2004), Kindle Edition; Daniel Schlozman, *When Movements Anchor Parties* (Princeton, NJ: Princeton University Press, 2015), Kindle Edition; Interview with Corey Robin conducted by Guido Girgenti, October 2019.

214 *this array of groups and movements*: Jonathan M. Smucker, *F*ckers at the Top:*

A Practical Guide to Overthrowing America's Ruling Class (Washington, DC: Strong Arm Press, 2020 [forthcoming]).

216 *four full years after the economic crash of 1929*: William E. Leuchtenburg, *Franklin D. Roosevelt and the New Deal: 1932–1940* (New York: Harper Perennial, 2009), 1–3.

217 *President Hoover held firm to his party's beliefs*: Sundquist, *Dynamics of the Party System*, 204.

217 *"Economic depression cannot be cured by legislative action"*: Herbert Hoover, *The Memoirs of Herbert Hoover, Vol. 3: The Great Depression, 1929–1941* (New York: Macmillan, 1952), 429–30.

217 *A new, broad coalition had come into power*: Ira Katznelson, *Fear Itself: The New Deal and the Origins of Our Time* (New York: Liveright, 2014), Kindle Edition, 22–24; Jean Edward Smith, *FDR* (New York: Random House, 2007), 374.

217 *"because the rulers of the exchange of mankind's goods have failed"*: Quoted in H. W. Brands, *Traitor to His Class* (New York: Anchor, 2009), 283.

217 *"replace the old order of special privilege"*: Quoted in Stephen Skowronek, *Franklin Roosevelt and the Modern Presidency* (Cambridge: Cambridge University Press, 1992), 328.

218 *guarantor of security and industrial stability*: Sundquist, *Dynamics of the Party System*, 210; Leuchtenburg, *Franklin D. Roosevelt and the New Deal:1932–1940*, 41–62, 331; Arthur M. Schlesinger, *The Politics of Upheaval: 1935–1936, The Age of Roosevelt, Volume 3* (Boston and New York: Houghton Mifflin/Mariner, 2003), Kindle Edition, 179.

218 *FDR lacked a singular "transformative vision"*: Quoted in Skowronek, *Franklin Roosevelt and the Modern Presidency*, 332; Alan Brinkley, *The End of Reform* (New York: Knopf Doubleday Publishing Group, 2011), Kindle Edition, 38.

218 *when FDR sought to stimulate industrial production*: Leuchtenburg, *Franklin D. Roosevelt and the New Deal: 1932–1940*, 57.

218 *Corporations demanded protection*: Katznelson, *Fear Itself*, 241–42; Leuchtenburg, *Franklin D. Roosevelt and the New Deal: 1932–1940*, 56–57.

218 *the agency signaled a profoundly reconstructed economic order*: Skowronek, *Franklin Roosevelt and the Modern Presidency*, 322–58; Katznelson, *Fear Itself*, 231.

218 *two general strikes, shutting down San Francisco and Minneapolis*: Leuchtenburg, *Franklin D. Roosevelt and the New Deal: 1932–1940*, 112–17.

218 *an extraordinary wave of labor militancy*: Leuchtenburg, *Franklin D. Roosevelt and the New Deal: 1932–1940*, 112–17; "Analysis of Strikes in 1938," Bureau of Labor Statistics, May 1939, https://www.bls.gov/wsp/publications/annual-summaries/pdf/analysis-of-strikes-in-1938.pdf.

219 *swept an even bigger Democratic majority into Congress*: Leuchtenburg, *Franklin D. Roosevelt and the New Deal: 1932–1940*, 114–17.

219 *"Congress threatened to push him in a direction far more radical"*: Ibid., 117.

219 *"We've got to get everything we want"*: Quoted in Schlesinger, *The Politics of Upheaval:1935–1936, The Age of Roosevelt, Volume 3*, 265.

219 *Wagner introduced the National Labor Relations Act*: Leuchtenburg, *Franklin D. Roosevelt and the New Deal: 1932–1940*, 115, 150.

219 *Roosevelt proposed a Social Security Act that stalled in the House*: Schlesinger, *The Politics of Upheaval: 1935–1936, The Age of Roosevelt, Volume 3*, 211.

219 *Consumers complained of corporations artificially inflating prices*: Robert H. Zieger, *The CIO, 1935–1955* (Chapel Hill: University of North Carolina Press, 1997), Kindle Edition, 211.

219 *small businesses and big firms sparred over pricing and market share*: Brinkley, *The End of Reform*, 39.

219 *gutted the heart of the first New Deal*: Theda Skocpol, Kenneth Finegold, and Michael Goldfield, "Explaining New Deal Labor Policy," *American Political Science Review* 84, no. 4 (1990):1297–1315.

219 *progressives urged Roosevelt to oppose business*: Schlesinger, *The Politics of Upheaval: 1935–1936, The Age of Roosevelt, Volume 3*, 274; Leuchtenburg, *Franklin D. Roosevelt and the New Deal: 1932–1940*, 36.

220 *Six weeks after the NRA was invalidated*: Katznelson, *Fear Itself*, 257.

220 *New Dealers compromised with Jim Crow*: Ibid., 163.

220 *ripe for mass unionization*: Nelson Lichtenstein, *State of the Union* (Princeton, NJ: Princeton University Press, 2002), 43.

220 *Lewis hired dozens of communists and socialists*: Lichtenstein, *State of the Union*, 45; Schlozman, *When Movements Anchor Parties*, 51.

220 *an impressive $600,000*: Zieger, *The CIO, 1935–1955*.

220 *fusion of racial justice and labor rights*: Eric Schickler, *Racial Realignment: The Transformation of American Liberalism 1932–1965*, Princeton Studies in American Politics: Historical, International, and Comparative Perspectives (Princeton, NJ: Princeton University Press, 2016), Kindle Edition, 8.

220 *successful crusade to pass the Fair Labor Standards Act*: Brinkley, *The End of Reform*, 219.

221 *FDR soon appointed him to represent labor*: Zieger, *The CIO, 1935–1955*.

221 *Hillman believed that labor support for Roosevelt's wartime program was essential*: Nelson Lichtenstein, *Labor's War at Home* (Philadelphia, PA: Temple University Press, 2003).

221 *Lewis broke the no-strike pledge, mobilizing miners in protest*: Lichtenstein, *Labor's War at Home*; Zieger, *The CIO, 1935–1955*.

221 *"leaders more like Lewis than Hillman"*: Brinkley, *The End of Reform*, 225.

221 *"labor organizing . . . stimulated civil rights activism"*: Katznelson, *Fear Itself*, 398.

221 *a business class that FDR believed he needed on his side to win WWII*: Brinkley, *The End of Reform*, 143.

221 *limited the legacy of the New Deal*: Alan Brinkley, "The New Deal and the Idea of the State," in *The Rise and Fall of the New Deal Order, 1930–1980*, ed. Steve Fraser and Gary Gerstle (Princeton, NJ: Princeton University Press, 1989); Nelson Lichtenstein, "From Corporatism to Collective Bargain," in Fraser and Gerstle, *The Rise and Fall of the New Deal Order, 1930–1980*.

221 *the core of the New Deal consensus*: Kim Phillips-Fein, *Invisible Hands: The Businessmen's Crusade Against the New Deal* (New York: W. W. Norton & Company, 2010), Kindle Edition.

221 *some wealthy businessmen began plotting*: Kim Phillips-Fein, *Invisible Hands: The Businessmen's Crusade Against the New Deal* (New York: W. W. Norton, 2010), Kindle edition.

222 *the growth of the postwar economy affirmed that this consensus worked*: Thomas Ferguson, "Industrial Conflict and the Coming of the New Deal," in Fraser and Gerstle, *The Rise and Fall of the New Deal Order, 1930–1980*; David F. Weiman, "Imagining a World Without the New Deal," *Washington Post*, August 3, 2011, https://www.washingtonpost.com/opinions/imagining-a-world-without-the-new-deal/2011/08/03/gIQAtJoBBJ_story.html.

222 *New Deal alignment established a common sense so dominant*: Sundquist, *Dynamics of the Party System*, 335, 336; Skowronek, *Presidential Leadership in Political Time*; Katznelson, *Fear Itself*, 473, 474.

222 *"Should any political party attempt to abolish social security . . . they are stupid"*: Quoted in Jean Edward Smith, *Eisenhower: In War and Peace* (New York: Random House, 2012), xiv. Also see Dwight Eisenhower to Edgar Newton Eisenhower, November 8, 1954, in David Mikkelson, "President Eisenhower on Social Security," Snopes.com, July 21, 2015, https://www.snopes.com/fact-check/social-insecurity/.

222 *denounced mainstream Republicans*: Barry Goldwater, *The Conscience of a Conservative* (Shepherdsville, KY: Victor Publishing Co., 1960), 15.

222 *pioneering conservative realigners*: Sam Rosenfeld and Daniel Schlozman, "The Long New Right and the World It Made," working paper, January 2019, https://static1.squarespace.com/static/540f1546e4b0ca60699c8f73/t/5c3e694321c67c3d28e992ba/1547594053027/Long+New+Right+Jan+2019.pdf.

222 *expanded the social safety net the New Deal first created*: Ira Katznelson, "Was the Great Society a Lost Opportunity?," in Fraser and Gerstle, *The Rise and Fall of the New Deal Order, 1930–1980*.

223 *"identified civil rights as a critical front"*: Schickler, *Racial Realignment*, 285.

223 *percentage of Americans living in poverty was cut in half*: Gordon Fisher, "Estimates of the Poverty Population Under Current Official Definition for Years Before 1959," US Department of Health and Human Services, 1986 (see John Iceland, *Poverty in America: A Handbook* [Berkeley: University of California Press, 2012], 74, fig. 5.1); "Historical Poverty Tables: People and Families—1959 to 2018," United States Census Bureau, August 27, 2019, https://www.census.gov/data/tables/time-series/demo/income-poverty/historical-poverty-people.html.

223 *President Nixon governed within the limits of New Deal consensus*: Corey Robin, "The Triumph of the Shill," *n+1* 29 (Fall 2017).

223 *Bayard Rustin actively worked to push Southern Democrats*: Paul Heideman, "It's Their Party," *Jacobin*, February 4, 2016.

224 *Martin Luther King tried to organize impoverished Americans of every race*: Paul Heideman, "It's Their Party," *Jacobin*, February 4, 2016; Schickler, *Racial Realignment*; Bayard Rustin, "From Protest to Politics: The Future of the Civil Rights Movement," *Commentary*, February 1965.

224 *labor-civil rights coalition at the helm of the Democratic Party was never fully realized*: Ibid, Heideman.

224 *Democrats steadily lost the South without solidifying a new majority themselves*: Sundquist, *Dynamics of the Party System*, 289, 345.

224 *"very much like the New Dealers of 40 years ago"*: Quoted in Schlozman, *When Movements Anchor Parties*, 82.

224 *"achieve political power through the coalition"*: Quoted in Rosenfeld and Schlozman, "The Long New Right and the World It Made," 7.

224 *he co-founded the Heritage Foundation*: Ibid.

224 *"the long-held conservative view that government is too powerful"*: Quoted in Schlozman, *When Movements Anchor Parties*, 92.

224 *"there's a moral majority"*: Quoted in David Grann, "Robespierre of the

Right," *New Republic*, October 27, 1997, https://newrepublic.com/article/61338/robespierre-the-right.

225 *"need not mean either compromise politics or party politics"*: Quoted in Rosenfeld and Schlozman, "The Long New Right and the World It Made."

225 *Schlafly built the Eagle Forum*: Donald T. Critchlow, *Phyllis Schlafly and Grassroots Conservatism: A Woman's Crusade* (Princeton, NJ: Princeton University Press, 2008), Kindle Edition, 221.

225 *further link abortion, antifeminism, and school prayer to their free market ideology*: Ibid., 262; Mark Depue, "Interview with Phyllis Schlafly," Abraham Lincoln Presidential Library Oral History Program, January 5, 2011, //www2.illinois.gov/alplm/library/collections/OralHistory/illinoisstatecraft/era/Documents/SchlaflyPhyllis/Schlafly_Phy_4FNL.pdf.

225 *a Women's Policy Board helmed by two pro-ERA activists*: Critchlow, *Phyllis Schlafly and Grassroots Conservatism*, 274.

225 *Schlafly sent an "emergency telegram"*: Ibid.

226 *increasingly under attack by a hostile business class*: Phillips-Fein, *Invisible Hands*.

226 *Corporate profits were in decline*: David Harvey, *A Brief History of Neoliberalism* (New York: Oxford University Press, 2007).

226 *CEOs rushed to find new allies*: Paul Heideman, "It's Their Party," *Jacobin*, February 4, 2016; Jacob S. Hacker and Paul Pierson, *Winner-Take-All Politics: How Washington Made the Rich Richer—and Turned Its Back on the Middle Class* (New York: Simon & Schuster, 2010).

226 *"an era ending—and a new one beginning"*: Quoted in Sundquist, *Dynamics of the Party System*, 1.

226 *"dominant order of ideas, public policies, and political alliances"*: Fraser and Gerstle, *The Rise and Fall of the New Deal Order, 1930–1980.*

227 *a thousand-page neoliberal policy manual, entitled* Mandate for Leadership: Daniel Stedman Jones, *Masters of the Universe: Hayek, Friedman, and the Birth of Neoliberal Politics* (Princeton, NJ: Princeton University Press, 2012), Kindle Edition,164.

227 *Heritage later boasted that Reagan implemented nearly two-thirds of the manual's policies*: Ibid.

227 *Reagan fired over 11,000 striking air traffic controllers*: Hacker and Pierson, *Winner-Take-All Politics*, 58–59.

227 *reported violations of the Wagner Act skyrocketed*: Ibid.

227 *the number of Americans living below the poverty line rose*: AP, "Census Bureau

Reports 1980 Poverty Statistics," *New York Times*, August 22, 1982, https://www.nytimes.com/1982/08/22/us/census-bureau-reports-1980-poverty-statistics.html; US Bureau of the Census, Current Population Reports, Series P-60, No. 171, Poverty in the United States: 1988 and 1998.

227 *39 percent of the nation's wealth*: Eduardo Porter, "Incomes Grew After Past Tax Cuts, but Guess Whose," *New York Times*, December 26, 2017, https://www.nytimes.com/2017/12/26/business/economy/tax-cuts-incomes.html; Josh Bivens, "The Top 1 Percent's Share of Income from Wealth Has Been Rising for Decades," Economic Policy Institute, April 23, 2014, https://www.epi.org/publication/top-1-percents-share-income-wealth-rising/.

227 *Reaganites removed New Deal protections that safeguarded families*: Paul Krugman, "Reagan Did It," *New York Times*, May 31, 2009, https://www.nytimes.com/2009/06/01/opinion/01krugman.html.

228 *from $33 million to $1,042 million . . . from $86 to $1,026 million*: Michelle Alexander, *The New Jim Crow* (New York: The New Press, 2010), 49.

228 *US prison population doubled from 315,974 to 739,980*: "Criminal Justice Facts," The Sentencing Project, n.d., accessed March 2020, https://www.sentencingproject.org/criminal-justice-facts/.

228 *to identify and repeal "any remaining discriminatory laws" against women*: Critchlow, *Phyllis Schlafly and Grassroots Conservatism*, 280.

228 *Schlafly opposed this project, pushing the administration*: Ibid.

228 *"We do not have a separate social agenda"*: Ibid., 273.

228 *his nomination of moderate Sandra Day O'Connor to the Supreme Court*: Ibid.

228 *infuriated pro-lifers*: Leslie Bennetts, "Antiabortion Forces in Disarray Less Than a Year After Victories in Elections," *New York Times,* September 22, 1981, https://www.nytimes.com/1981/09/22/us/antiabortion-forces-in-disarray-less-than-a-year-after-victories-in.html.

228 *"We are a movement in disarray"*: Leslie Bennetts, "Antiabortion Forces in Disarray Less Than a Year After Victories in Election."

229 *"the social issues aren't big in the Country Clubs"*: Quoted in Rosenfeld and Schlozman, "The Long New Right and the World It Made," 49.

229 *successfully "reshaped the national agenda"*: Hedrick Smith, "Reagan's Effort to Change Course of Government," *New York Times*, October 23, 1984, https://www.nytimes.com/1984/10/23/us/reagan-s-effort-to-change-course-of-government.html.

229 *"it always seemed more like the Great Rediscovery"*: Ronald Reagan, "Transcript of Reagan's Farewell Address to American People," *New York Times*,

January 12, 1989, https://www.nytimes.com/1989/01/12/news/transcript-of-reagan-s-farewell-address-to-american-people.html.

229 *"America is still working its way through the Reagan era"*: Skowronek, *Presidential Leadership in Political Time: Reprise and Reappraisal*, 4.

229 *widely read 1982 op-ed entitled "A Neo-Liberal's Manifesto"*: Charles Peters, "A Neo-Liberal's Manifesto," *Washington Post*, September 5, 1982, https://www.washingtonpost.com/archive/opinions/1982/09/05/a-neo-liberals-manifesto/21cf41ca-e60e-404e-9a66-124592c9f70d/.

230 *Governor Bill Clinton oversaw the execution of Ricky Ray Rector*: Michelle Alexander, "Why Hillary Clinton Doesn't Deserve the Black Vote," *The Nation*, February 10, 2016, https://www.thenation.com/article/archive/hillary-clinton-does-not-deserve-black-peoples-votes/.

230 *"No one can say I'm soft on crime"*: Ibid.

230 *US prison population rose nearly 60 percent*: Ed Pilkington, "Bill Clinton: Mass Incarceration on My Watch 'Put too Many People in Prison,'" *The Guardian*, April 13, 2015, https://www.theguardian.com/us-news/2015/apr/28/bill-clinton-calls-for-end-mass-incarceration.

230 *dismantled the Aid to Families with Dependent Children*: Vann R. Newkirk II, "The Real Lessons from Bill Clinton's Welfare Reform," *The Atlantic*, February 5, 2018, https://www.theatlantic.com/politics/archive/2018/02/welfare-reform-tanf-medicaid-food-stamps/552299/; Ife Floyd, Maritzelena Chirinos, and Nick McFaden, "Cash Assistance Should Reach Millions More Families," Center on Budget and Policy Priorities, March 4, 2020, https://www.cbpp.org/research/family-income-support/tanf-reaching-few-poor-families.

230 *"broader, deeper and more dangerous"*: Barry Ritholtz, "Repeal of Glass-Steagall: Not a Cause but a Multiplier," *Washington Post*, August 4, 2012, https://www.washingtonpost.com/repeal-of-glass-steagall-not-a-cause-but-a-multiplier/2012/08/02/gJQAuvvRXX_story.html.

231 *a "New Deal"*: Quoted in William E. Leuchtenburg, *In the Shadow of FDR* (Ithaca, NY: Cornell University Press, 2011), Kindle Edition.

231 *Obama praised Clinton's Third Way*: Barack Obama, *The Audacity of Hope* (New York: Crown, 2007), Kindle Edition, 34.

231 *"if [it's] right for the times then we're going to apply it"*: Leuchtenburg, *In the Shadow of FDR*.

231 *The goals of the stimulus*: Reed Hundt, *A Crisis Wasted: Barack Obama's Defining Decisions* (New York: Simon & Schuster/RosettaBooks, 2019), Kindle Edition, 142–45.

232 *Summers made clear he would not*: Reed Hundt, *A Crisis Wasted: Barack Obama's*

Defining Decisions (New York: Simon & Schuster/RosettaBooks, 2019), Kindle Edition, 108; Noam Scheiber, "EXCLUSIVE: The Memo that Larry Summers Didn't Want Obama to See," *New Republic*, February 22, 2012. https://newrepublic.com/article/100961/memo-larry-summers-obama.

232 *"the problem is that you are talking about creating more debt"*: Hundt, *A Crisis Wasted*, 123.

232 *Treasury Secretary Timothy Geithner set guidelines*: Ibid., 105.

232 *With the moonshots set aside*: Alex Seitz-Wald, "Sen. Shelby Falsely Claims Stimulus Package's $288 Billion in Tax Cuts Were 'More Taxes,'" ThinkProgress, September 20, 2011, https://thinkprogress.org/sen-shelby-falsely-claims-stimulus-packages-288-billion-in-tax-cuts-were-more-taxes-19df1d8866e0/; "Estimated Impact of the American Recovery and Reinvestment Act on Employment and Economic Impact from October 2011 Through December 2011," Congressional Budget Office, February 2012, https://www.cbo.gov/sites/default/files/cbofiles/attachments/02-22-ARRA.pdf; Brad Plumer, " A Closer Look at Obama's '$90 Billion for Green Jobs,'" *Washington Post*, October 4, 2012, https://www.washingtonpost.com/news/wonk/wp/2012/10/04/a-closer-look-at-obamas-90-billion-for-clean-energy/.

232 *Over 30 million Americans would lose their jobs in the course of the Great Recession*: Arne L. Kalleberg and Till M. von Wachter, "The US Labor Market During and After the Great Recession: Continuities and Transformations," National Center for Biotechnology Information, April 2017, https://www.ncbi.nlm.nih.gov/pmc/articles/PMC5959048/.

232 *9.3 million Americans lost their homes*: Laura Kusisto, "Many Who Lost Homes to Foreclosure in Last Decade Won't Return—NAR," *Wall Street Journal*, April 20, 2015, https://www.wsj.com/articles/many-who-lost-homes-to-foreclosure-in-last-decade-wont-return-nar-1429548640.

233 *National Commission on Fiscal Responsibility and Reform*: Jackie Calmes, "Obama's Deficit Dilemma," *New York Times*, February 27, 2012, https://www.nytimes.com/2012/02/27/us/politics/obamas-unacknowledged-debt-to-bowles-simpson-plan.html.

234 *less than three years after Wall Street crashed the economy*: Dave Boyer, "Obama Calls for 'Shared Sacrifice,'" *Washington Times*, August 17, 2011, https://www.washingtontimes.com/news/2011/aug/17/prepping-debt-plan-obama-calls-shared-sacrifice/.

234 *"The door is closing on the Reagan-Clinton era"*: Peter Beinart, "The Rise of

the New New Left," *Daily Beast*, July 11, 2017, https://www.thedailybeast.com/the-rise-of-the-new-new-left.

235 *Trump might "realign" the parties*: Lee Drutman, "Donald Trump's Candidacy Is Going to Realign the Political Parties," *Vox*, March 1, 2016, https://www.vox.com/polyarchy/2016/3/1/11139054/trump-party-realignment.

235 *"the end of the Reagan era"*: Julia Azari, "Trump's Presidency Signals the End of the Reagan Era," *Vox*, December 1, 2016, https://www.vox.com/mischiefs-of-faction/2016/12/1/13794680/trump-presidency-reagan-era-end.

235 *Their vision is exhausted*: Interview with Corey Robin conducted by Guido Girgenti, October 2019; Corey Robin with Chris Hayes interview, May 15, 2018, *Why Is This Happening?* (podcast), https://www.nbcnews.com/think/opinion/corey-robin-conservative-movement-podcast-transcript-ncna874126.

236 *Biden's platform has moved far leftward*: David Brooks, "Why Sanders Will Probably Win the Nomination," *New York Times*, February 20, 2020, https://www.nytimes.com/2020/02/20/opinion/bernie-sanders-win-2020.html; Jonathan Chait, "Joe Biden's Platform Is More Progressive Than You Think," *New York Magazine*, March 12, 2020, https://nymag.com/intelligencer/2020/03/joe-biden-platform-progressive-health-care-climate-taxes.html.

236 *a leftward-moving electorate*: Matthew Yglesias, "Public Support for Left-Wing Policymaking Has Reached a 60-Year High," *Vox*, June 7, 2019, https://www.vox.com/2019/6/7/18656441/policy-mood-liberal-stimson.

236 *Reagan handily won the youth vote*: Steven V. Roberts, "Younger Voters Tending to Give Reagan Support," *New York Times*, October 16, 1984, https://www.nytimes.com/1984/10/16/us/younger-voters-tending-to-give-reagan-support.html.

237 *"shared responsibility is not diminished responsibility"*: Quoted in Ross Douthat, "The Democrats Have a Culture Problem," March 5, 2019, *New York Times*, https://www.nytimes.com/2019/03/05/opinion/democrats-liberals-socialists-cultural-left.html; also see Brad DeLong (@delong), Twitter, February 25, 2019, 5:49 p.m., https://twitter.com/delong/status/1100166024572239873.

237 *a former vice president at the Democratic Leadership Council*: Ed Kilgore, "A New Role for Democratic Centrists: Helping the Left Win," *New York Magazine*, March 5, 2019, https://nymag.com/intelligencer/2019/03/a-new-role-for-democratic-centrists-helping-the-left-win.html.

237 *"fine-tuning a grossly unjust economy and a corrupt political system"*: George Packer, "Is America Undergoing a Political Realignment?," *The Atlantic*, April 8, 2019, https://www.theatlantic.com/ideas/archive/2019/04/will-2020-bring-realignment-left/586624/.

237 *"rapid and far-reaching transitions"*: V. Masson-Delmotte et al., eds., "Summary for Policymakers," in *Global Warming of 1.5°C: An IPCC Special Report on the Impacts of Global Warming of 1.5°C Above Pre-industrial Levels and Related Global Greenhouse Gas Emission Pathways, in the Context of Strengthening the Global Response to the Threat of Climate Change, Sustainable Development, and Efforts to Eradicate Poverty*," IPCC, 2018, https://www.ipcc.ch/sr15/chapter/spm/.

237 *"Global greenhouse gas emissions must begin falling by 7.6 percent each year"*: Brady Dennis, "In Bleak Report, U.N. Says Drastic Action Is Only Way to Avoid Worst Effects of Climate Change," *Washington Post*, November 26, 2019, https://www.washingtonpost.com/climate-environment/2019/11/26/bleak-report-un-says-drastic-action-is-only-way-avoid-worst-impacts-climate-change/.

238 *"Borders are the environment's greatest ally"*: Aude Mazoue, "Le Pen's National Rally Goes Green in Bid for European Election Votes," France 24, April 20, 2019, https://www.france24.com/en/20190420-le-pen-national-rally-front-environment-european-elections-france.

238 *mustering a "second wind"*: Sanford Levinson and Jack M. Balkin, *Democracy and Dysfunction* (Chicago: University of Chicago Press, 2019), Kindle Edition, 192.

CHAPTER 15: *FROM PROTEST TO PRIMARIES: THE MOVEMENT IN THE DEMOCRATIC PARTY* / ALEXANDRA ROJAS AND WALEED SHAHID

246 *"Too many members worried"*: Jay Newton-Small, "5 Reasons Immigration Reform Is Going Nowhere Fast," *Time*, February 6, 2014, https://time.com/4999/immigration-reform-house-senate-boehner/.

247 *the challenger was outspent nearly 25 to 1*: Matthew Yglesias, "Cantor outspent Brat 25:1," *Vox*, June 10, 2014, https://www.vox.com/2014/6/10/5798674/cantor-outspent-brat-25-1.

249 *"people try to accuse us of going too far left"*: Sydney Ember, "Ocasio-Cortez and Sanders Star in Their Own Iowa Buddy Movie," *New York Times*, November 10, 2019, https://www.nytimes.com/2019/11/10/us/politics/bernie-sanders-aoc-iowa.html.

250 *"Despite the widespread belief"*: Nolan McCarty, "What We Know and Don't Know About Our Polarized Politics," *Washington Post*, January 8, 2014,

https://www.washingtonpost.com/news/monkey-cage/wp/2014/01/08/what-we-know-and-dont-know-about-our-polarized-politics/?arc404=true.

250 *Feeling the intensity of the Tea Party*: Derek Thompson, "Reality Check: Obama Cuts Social Security and Medicare by Much More Than the GOP," *The Atlantic*, April 11, 2013, https://www.theatlantic.com/business/archive/2013/04/reality-check-obama-cuts-social-security-and-medicare-by-much-more-than-the-gop/274919/.

252 *"He who can make the nominations is the owner of the party"*: E. E. Schattschneider, *Party Government: American Government in Action* (Abingdon, UK: Routledge), 64.

253 *And by putting that vision on the ballot*: Cedric de Leon, Manali Desai, and Cihan Tuğal, *Building Blocs: How Parties Organize Society* (Stanford, CA: Stanford University Press, 2015), 91.

254 *"I'm not running 'from the left'*: Alexandria Ocasio-Cortez (@AOC), Twitter, July 3, 2018, 11:41 a.m., https://twitter.com/aoc/status/1014172302777507847?lang=en.

256 *"develop new ideas, refine them into workable policies"*: Daniel DiSalvo, *Engines of Change: Party Factions in American Politics, 1868–2010* (New York: Oxford University Press, 2012), xii.

257 *"They deploy resources to favored candidates"*: Daniel Schlozman, *When Movements Anchor Parties: Electoral Alignments in American History* (Princeton, NJ: Princeton University Press, 2015), 21.

259 *"diluted, milk and water gruel"*: Thaddeus Stevens, Cong. Globe, 37th Cong., 2d Sess. (1862), 1154, https://memory.loc.gov/cgi-bin/ampage.

259 *"are nearer to me than the other side"*: Eric Foner, *Fiery Trial* (New York: W. W. Norton & Company, 2010), 544.

260 *"It turned me blind . . . seeing the people there"*: Ibid., 189.

260 *"the demographic and political majority that their generation will become"*: Barbara Ransby, "'The Squad' Is the Future of Politics," *New York Times*, August 8, 2019, https://www.nytimes.com/2019/08/08/opinion/the-squad-democrats.html.

CHAPTER 16: *REVIVING LABOR, IN NEW DEALS OLD AND GREEN* / BOB MASTER

264 *individualism of the Gilded Age*: Eric Foner, *The Story of American Freedom* (New York: W. W. Norton & Company, 1998), 195–201.

264 *"vision of social reconstruction"*: Michael Denning, *The Cultural Front* (London and New York: Verso, 1997), 3–7.

266 *as UAW president Doug Fraser dubbed it back in 1978*: Jefferson Cowie, " 'A One-Sided Class War': Rethinking Doug Fraser's 1978 Resignation from the Labor-Management Group," *Labor History* 44, no. 3 (2003): 307–14.

266 *mass media would all be weaponized against the aging New Deal consensus*: Lewis F. Powell, "Confidential Memorandum: Attack on American Free Enterprise System," August 23, 1971, accessed at https://www.greenpeace.org/usa/democracy/the-lewis-powell-memo-a-corporate-blueprint-to-dominate-democracy/.

267 *social programs upon which working people depend*: Nancy MacLean, *Democracy in Chains: The Deep History of the Radical Right's Stealth Plan for America* (New York: Penguin Books, 2017); Ian Haney López, *Dog Whistle Politics: How Coded Racial Appeals Have Reinvented Racism and Wrecked the Middle Class* (New York: Oxford University Press, 2014).

267 *little space in the public imagination for unions*: Perry Anderson, "Renewals," *New Left Review* 1, no. 1 (January–February 2000): 13.

268 *endows unions with social, political, and moral authority and purpose*: Nelson Lichtenstein, *State of the Union* (Princeton, NJ: Princeton University Press, 2002), 18–19, 43.

269 *"everything—now or never"*: William Leuchtenburg, *Franklin D. Roosevelt and the New Deal: 1932–1940* (New York: Harper & Row, 1963), 116–17.

269 "*most radical piece of legislation ever enacted by the United States Congress*": Karl Klare, "Judicial Deradicalization of the Wagner Act and the Origins of Modern Legal Consciousness, 1937–1941," *Minnesota Law Review* 62 (1977–1978): 265, 285.

270 *"unprecedented militance and tactical boldness"*: Steve Fraser, "The 'Labor Question,' " in *The Rise and Fall of the New Deal Order, 1930–1980*, ed. Steve Fraser and Gary Gerstle (Princeton, NJ: Princeton University Press, 1989), 67.

270 *"power of the emergent Roosevelt coalition"*: Nelson Lichtenstein, *Walter Reuther: The Most Dangerous Man in Detroit* (New York: Basic Books, 1995), 61–63.

270 *" 'Now get a New Deal in the shop' "*: Lichtenstein, *State of the Union,* 48.

271 *" 'my boss is a son of a bitch' "*: Ibid., 46.

271 *"one of the greatest chapters in the historic struggle for human liberties in this country"*: David Montgomery, "American Workers and the New Deal Formula," in David Montgomery, *Workers' Control in America* (Cambridge: Cambridge University Press, 1980), 165.

272 "*lost credibility as inequality has widened*": Noam Scheiber, "Candidates

Grow Bolder on Labor, and Not Just Bernie Sanders," *New York Times,* October 12, 2019, https://www.nytimes.com/2019/10/11/business/economy/democratic-candidates-labor-unions.html.

274 *congressional power brokers of the Bourbon South*: Ira Katznelson, *Fear Itself: The New Deal and the Origins of Our Time* (New York: Liveright, 2013).

274 *slash government spending by 25 percent*: Leuchtenburg, *Franklin D. Roosevelt and the New Deal: 1932–1940*, 11.

274 *searing Arizona desert in 1917*: Adam Cohen, *Nothing to Fear: FDR's Inner Circle and the Hundred Days That Created Modern America* (New York: Penguin Books, 2009), 88–106; Leuchtenburg, *Franklin D. Roosevelt and the New Deal: 1932–1940,* 41–47.

275 *capitalist system's answer to that question*: Fraser, "The 'Labor Question,'" 55.

275 *the climate crisis is the defining issue of* our *time*: Tim Flannery, "Australia: The Fires and Our Future," *New York Review of Books*, January 16, 2020, https://www.nybooks.com/daily/2020/01/16/australia-the-fires-and-our-future/.

276 *"limiting membership to 'whites' or 'Caucasians'"*: Judith Stepan-Norris and Maurice Zeitlin, *Left Out: Reds and America's Industrial Unions* (Cambridge: Cambridge University Press, 2003), fn, 232.

276 *institutions that arose during the New Deal era*: Lichtenstein, *State of the Union,* 39–41, 64–74.

INDEX

Note: page numbers in italics refer to images.